DEFENSE, SECURITY AND STRATEGIES

CIVIL RESERVE AIR FLEET

BACKGROUND AND ANALYSES

DEFENSE, SECURITY AND STRATEGIES

Additional books in this series can be found on Nova's website under the Series tab.

Additional e-books in this series can be found on Nova's website under the e-book tab.

DEFENSE, SECURITY AND STRATEGIES

CIVIL RESERVE AIR FLEET

BACKGROUND AND ANALYSES

SOREN M. JONSSON
EDITOR

nova publishers
New York

For permission to use material from this book please contact us:
Telephone 631-231-7269; Fax 631-231-8175
Web Site: http://www.novapublishers.com

NOTICE TO THE READER

Additional color graphics may be available in the e-book version of this book.

Library of Congress Cataloging-in-Publication Data

ISBN: 978-1-62808-782-6

Published by Nova Science Publishers, Inc. †New York

CONTENTS

PREFACE

This book provides background and analyses on the Civil Reserve Air Fleet (CRAF) which was created by executive order in 1951. As a result, the Departments of Commerce (DOC) and Defense (DOD) formulated a contingency plan to meet the nation's airlift needs in times of crisis. When the Department of Transportation (DOT) was created, it assumed DOC's role in the CRAF program, and today, DOD and DOT work together to manage the CRAF program. The CRAF supports DOD airlift requirements in emergencies when the need for airlift exceeds the capability of the military aircraft fleet. All CRAF participants must be U.S. carriers fully certified by the Federal Aviation Administration, and meet the stringent standards of Federal Aviation Regulations pertaining to commercial airlines. The CRAF has three main segments: international, national, and aeromedical evacuation. The international segment is further divided into the long-range and short-range sections and the national segment into the domestic and Alaskan sections. Assignment of aircraft to a segment depends on the nature of the requirement and the performance characteristics needed.

Chapter 1 – The Civil Reserve Air Fleet (CRAF) was created by executive order in 1951. As a result, the Departments of Commerce (DOC) and Defense (DOD) formulated a contingency plan to meet the nation's airlift needs in times of crisis. When the Department of Transportation (DOT) was created, it assumed DOC's role in the CRAF program, and today, DOD and DOT work together to manage the CRAF program. The CRAF supports DOD airlift requirements in emergencies when the need for airlift exceeds the capability of the military aircraft fleet. All CRAF participants must be U.S. carriers fully certified by the Federal Aviation Administration, and meet the stringent standards of Federal Aviation Regulations pertaining to commercial airlines.

The CRAF has three main segments: international, national, and aeromedical evacuation. The international segment is further divided into the long-range and short-range sections and the national segment into the domestic and Alaskan sections. Assignment of aircraft to a segment depends on the nature of the requirement and the performance characteristics needed. The commercial airlines contractually pledge aircraft to the various segments of CRAF, ready for activation when needed. To provide incentives for civil carriers to commit aircraft to the CRAF program and to assure the United States of adequate airlift reserves, the government makes peacetime airlift business available to civilian airlines that obligate aircraft to the CRAF. DOD offers business through the International Airlift Services. CRAF presents benefits and opportunities for both DOD and U.S. airlines. By all accounts it appears to be a symbiotic relationship. Yet, as circumstances change, pressures and diverging interests may emerge that could bring changes to CRAF. A number of factors may be considered when examining the future size, character and role of CRAF. These factors include cost, other potential government / commercial arrangements, potential change in DOD requirement for CRAF, and industrial base or financial assistance to U.S. air carriers.

Chapter 2 – Report on Congressional Budget Office on Issues Regarding the Current and Future Use of the Civil Reserve Air Fleet, dated October 2007.

Chapter 3 – Report on United States Government Accountability Office on Military Airlift: DOD Should Take Steps to Strengthen Management of the Civil Reserve Air Fleet Program, dated September 2009.

Chapter 4 – Report on United States Government Accountability Office on Military Airlift: DOD Needs to Take Steps to Manage Workload Distributed to the Civil Reserve Air Fleet, dated June 2013.

In: Civil Reserve Air Fleet
Editor: Soren M. Jonsson

ISBN: 978-1-62808-782-6

Chapter 1

CIVIL RESERVE AIR FLEET (CRAF)*

William Knight and Christopher Bolkcom

SUMMARY

The Civil Reserve Air Fleet (CRAF) was created by executive order in 1951. As a result, the Departments of Commerce (DOC) and Defense (DOD) formulated a contingency plan to meet the nation's airlift needs in times of crisis. When the Department of Transportation (DOT) was created, it assumed DOC's role in the CRAF program, and today, DOD and DOT work together to manage the CRAF program.

The CRAF supports DOD airlift requirements in emergencies when the need for airlift exceeds the capability of the military aircraft fleet. All CRAF participants must be U.S. carriers fully certified by the Federal Aviation Administration, and meet the stringent standards of Federal Aviation Regulations pertaining to commercial airlines.

The CRAF has three main segments: international, national, and aeromedical evacuation. The international segment is further divided into the long-range and short-range sections and the national segment into the domestic and Alaskan sections. Assignment of aircraft to a segment depends on the nature of the requirement and the performance characteristics needed.

The commercial airlines contractually pledge aircraft to the various segments of CRAF, ready for activation when needed. To provide

* This is an edited, reformatted and augmented version of a Congressional Research Service publication, CRS Report for Congress RL33692, prepared for Members and Committees of Congress, from www.crs.gov, dated April 25, 2008.

incentives for civil carriers to commit aircraft to the CRAF program and to assure the United States of adequate airlift reserves, the government makes peacetime airlift business available to civilian airlines that obligate aircraft to the CRAF. DOD offers business through the International Airlift Services.

CRAF presents benefits and opportunities for both DOD and U.S. airlines. By all accounts it appears to be a symbiotic relationship. Yet, as circumstances change, pressures and diverging interests may emerge that could bring changes to CRAF. A number of factors may be considered when examining the future size, character and role of CRAF. These factors include cost, other potential government / commercial arrangements, potential change in DOD requirement for CRAF, and industrial base or financial assistance to U.S. air carriers.

Introduction

The Civil Reserve Air Fleet (CRAF) was created by President Truman in 1951.[1] As a result, the Departments of Commerce (DOC) and Defense (DOD) formulated a contingency plan to meet the nation's airlift needs in times of crisis. When the Department of Transportation (DOT) was created, it assumed DOC's role in the CRAF program, and today, DOD and DOT work together to manage the CRAF program.[2] This report provides background, analyzes current issues, and summarizes recent legislation for the CRAF.

Background

The CRAF supports DOD airlift requirements in emergencies when the need for airlift exceeds the capacity of DOD's organic airlift fleet. While DOD strategic airlift aircraft are designed to carry outsized and oversized cargo,[3] CRAF air carriers are primarily expected to transport passengers and cargo pallets. All CRAF participants must be U.S. carriers fully certified by the Federal Aviation Administration (FAA), and meet the stringent standards of the Federal Aviation Regulations pertaining to commercial airlines (Part 121). To join CRAF, a carrier must commit at least 30% of its CRAF-capable passenger fleet, and 15% of its CRAF-capable cargo fleet. Aircraft committed must be U.S. registered and air carriers must also commit and maintain at least four complete crews for each aircraft in CRAF.[4]

Air Mobility Command (AMC) analysts implement a number of surveillance initiatives to monitor the carrier's safety record, operations and maintenance status, contract performance, financial condition and management initiatives, summarizing significant trends in a comprehensive review every six months. These initiatives are supplemented by an open flow of information on all contract carriers between AMC and the FAA through established liaison officers.[5]

CRAF Structure

The CRAF has three main segments: international, national, and aeromedical evacuation. The international segment is further divided into the long-range and short-range sections, while the national segment is divided into the domestic and Alaskan sections. Assignment of aircraft to a segment depends on the nature of the requirement and the aircraft performance characteristics needed.[6]

International Segment. The long-range international section consists commercial airliners capable of transoceanic operations. Medium-sized passenger and cargo aircraft make up the short-range international section supporting near offshore airlift requirements.

National Segment. The aircraft in the Alaskan section provide airlift within U.S. Pacific Command's area of responsibility, specific to Alaska needs. The domestic section is designed to satisfy increased DOD airlift requirements in the United States during an emergency.

Aeromedical Evacuation Segment. The aeromedical evacuation segment assists in the evacuation of casualties from operational theaters to hospitals in the continental United States. Kits containing litter stanchions, litters, and other aeromedical equipment are used to convert civil Boeing 767 passenger aircraft into air ambulances.

Contractual Relationship

The airlines contractually pledge aircraft to the various segments of CRAF, ready for activation when needed. To provide incentives for civil carriers to commit aircraft to the CRAF program and to assure the United States of adequate airlift reserves, the government makes peacetime airlift business available to civilian airlines that obligate aircraft to the CRAF through the International Airlift Services.[7]

For FY2007, the guaranteed portion of DOD's CRAF contract was $379 million, while AMC expected to award $2.1 billion in additional business that were not guaranteed.[8] The Air Force announced $2.2 billion in CRAF contracts had been let in FY2005.[9] DOD let contracts worth $3.8 billion between FY1998 and FY2002:[10] 1998, $646 million; 1999, $710 million; 2000, $629 million; 2001, $572 million; and 2002, $1,280 million.

Activation

Three stages of incremental activation allow for tailoring an airlift force suitable for the contingency at hand.[11] The stages of activation are as follows:

- Stage I — minor regional crises.
- Stage II — major theater war.
- Stage III — periods of national mobilization.

The commander, U.S. Transportation Command (TRANSCOM), with approval of the Secretary of Defense, is the activation authority for all three stages of CRAF. During a crisis, if the Air Force Air Mobility Command (AMC) has a need for additional aircraft, it would request the TRANSCOM commander to take steps to activate the appropriate CRAF stage. Each stage of the CRAF activation is only used to the extent necessary to provide the amount of civil augmentation airlift needed by DOD. When notified of call-up, the carrier response time to have its aircraft ready for a CRAF mission is 24 to 48 hours after the mission is assigned by AMC. The air carriers continue to operate and maintain the aircraft with their resources; however, AMC controls the aircraft missions.[12]

CRAF has been formally activated on two separate occasions over the program's 57-year history. The first instance occurred for Operations Desert Shield/Storm from August 18, 1990, through May 24, 1991, and included long-range international passenger and cargo segments up to Stage II. During Operation Desert Storm, CRAF airlines executed 5,460 missions transporting 726,000 passengers and 230,000 tons of cargo at a cost of $1.4 billion. The second activation, during Operation Iraqi Freedom, lasted from February 8, 2003 through June 18, 2003, and included the long-range international passenger segment up to Stage 1 — long-range cargo requirements were met organically or with voluntary commercial contracts.[13]

CRAF Membership

As of March 2008, 35 carriers with 1,262 aircraft were enrolled in the CRAF. This includes 1,172 aircraft in the international segment (905 long-range and 267 short-range), 40 aircraft in the national segment, and 50 aircraft in the aeromedical evacuation segments.[14] *Table 1* summarizes current CRAF members:

Table 1. Airlines Participating in the Civil Reserve Air Fleet

International Segment		
Long Range Section		Short Range Section
ABX Air	Kalitta Air	Alaska Airlines
Air Transport Intl.	Murray Air	American Airlines
American Airlines	North American Airlines	ATA Airlines[a]
ATA Airlines[a]	Northwest Airlines	Champion Air[b]
Arrow Air	Omni International	Delta Airlines
Atlas Air	Polar Air Cargo	DHL Airways
Continental Airlines	Ryan Intl. Airlines	Jet Blue Airways
Delta Air Lines	Southern Air	Lynden Air Cargo
DHL Airways	United Airlines	Miami Air International
Evergreen International	United Parcel Service	Northern Air Cargo
FEDEX Airlines	U.S. Airways	Northwest Airlines
Gemini Air Cargo	World Airways	Sun Country
Domestic Section		Alaska Section
Air Trans Airways	Midwest Airlines	Lynden Air Cargo
Frontier Airlines	Southwest Airlines	Northern Air Cargo
Delta Air Lines	United Airlines	U.S. Airways

Source: Department of Transportation which issues updates monthly.

[a] ATA Airlines discontinued operations and file for bankruptcy protection under Chapter 11 on April 2, 2008, resulting in the loss of 31 aircraft to CRAF. See online at [http://www.ata.com/about_ ata/ news/0080403.pdf].

[b] Champion Air plans to discontinue flight operations on May 31, 2008 resulting in the loss of 6 aircraft to CRAF. See online at http://www.championair.com/ PressRoom.asp?PressRelease ID=21].

ANALYSIS: POTENTIAL FUTURE OF CRAF

CRAF presents benefits and opportunities for both the DOD and the U.S. airline industry — by all accounts it appears to be a symbiotic relationship. Yet, as circumstances change, pressures and diverging interests may emerge that could bring changes to CRAF.

Cost Factors

The primary benefit that CRAF imparts to DOD is its relatively low cost when compared to procuring and maintaining a larger organic fleet. For example, a 1996 Government Accountability Office (GAO) report noted that CRAF "provides up to half of the nation's strategic airlift capability without the government having to buy additional aircraft, pay personnel costs, or maintain the aircraft during peacetime"[15] — all factors that remain relevant today.

While CRAF is relatively inexpensive, some may point out that commercial aircraft have operational limitations when compared to DOD's organic airlift fleet. For example, commercial aircraft cannot carry outsized cargo, conduct special missions such as airdrop, or support special operations forces. Also, commercial aircraft tend to congest airfields because of longer ground times resulting from a lack of roll on/roll off capability and reduced ramp maneuverability. Further, potential hostile fire effectively deters civilian crews from entering combat zones. However, commercial aircraft typically have longer range, and are optimized to efficiently transport passengers and cargo pallets.[16] GAO references the use of CRAF during Operation Desert Storm to illustrate CRAF's cost advantages:

> The use of CRAF aircraft during an activation is not free — DOD pays rates based on weighted average carrier costs — but the cost is minimal in comparison to the costs of acquiring and supporting aircraft, paying and training aircrew, and other expenses of maintaining standby military airlift capability. AMC paid the carriers about $1.5 billion for using their aircraft during the operation. Purchasing additional military aircraft to provide similar capability would cost from $15 to $50 billion, according to Air Force officials, depending on assumptions used for aircraft replacement cost.[17]

A RAND study (*Finding the Right Mix of Military and Civil Airlift, Issues and Implications*) also includes a discussion of the cost-effectiveness of CRAF:

> For a very small cost, the DOD has had on call a very substantial airlift capacity. Replacing CRAF's 1992 Stage II capability with military-style transports would have cost the DOD about $1 billion annually (1992 dollars) over the past several decades. Replacing the Stage III capability would have cost about $3 billion annually.[18]

The RAND analysis points out that to have adequate airlift for a major crisis, DOD maintains a military airlift fleet with a total capacity four to five times greater than the average daily use. Costs associated with acquiring and maintaining this excess airlift capability must be routinely incurred, even if the full capacity is rarely used.

As DOD's procurement and operations and maintenance accounts come under increasing pressure, it may appear attractive to increase the size of CRAF in lieu of procuring and operating a certain fraction of the Air Force strategic airlift fleet. Recent events may suggest that a growing use of commercial aircraft for every-day DOD needs is already in evidence. In January 2005, for example, it was reported that commercial airlines moved twice as many U.S. troops overseas as they moved in January 2004.[19]

Contracting with air carriers to commit their aircraft to wartime needs is cheaper, in a sense, than purchasing and operating additional Air Force cargo aircraft. However, CRAF is not free, and it costs more once activated. RAND points out that:

> Although holding reserve capacity in the CRAF is far more cost effective than holding the reserve in the military airlift fleet, the government has a financial incentive to use its own resources (for which it has already committed funds) in a crisis to the extent that they are conveniently available, rather than give additional business to CRAF carriers.[20]

Other Government / Commercial Arrangements

CRAF is not the only means by which DOD transports troops by civil aircraft. Through the General Services Administration (GSA), the U.S.

government negotiates and lets contracts to commercial airlines to fly government employees on official U.S. government business. Federal employees, including DOD civilian and military personnel, traveling on government business are obliged to fly with these contracted airlines at the official government rate. DOD also charters commercial aircraft to satisfy peacetime mobility needs.

In July 2006 the U.S. Central Command had initiated a pilot program — "Commercial and Government Air Program" — to enlist commercial air cargo carriers to deliver military supplies into Afghanistan and Iraq. The pilot program is hoped to deliver up to 20% of DOD cargo to the region and to save DOD approximately $9 million per month.[21] DOD hopes to dramatically reduce its flight costs by creating competition among carriers for the work, and by leveraging excess cargo capacity on regularly scheduled commercial flights. This trial program could be viewed as something of an alternative to CRAF, or an indication that more CRAF would be welcome.

Potential Changes in DOD's Airlift Requirements and Force Structure

The increased scope and pace of military operations following the terrorist attacks of September 11, 2001, have increased the Air Force's mobility needs and made commercial air carriers a more prominent component of this capability.[22] Potential changes in DOD's strategic airlift requirements and air mobility force structure may affect the CRAF program.[23]

DOD Airlift Requirements. DOD periodically examines the state of its current air mobility fleet and quantifies future airlift requirements to determine whether current force structure is sufficient to meet the President's national security strategy. DOD's most recent air mobility requirements study, Mobility Capability Study (MCS), was completed in December 2005. However, during congressional testimony, General Arthur L. Lichte, Commander of the Air Force's Air Mobility Command, pointed to changes that have occurred since the MCS was completed that include the increase of 92,000 ground forces, the repositioning of DOD force structure overseas, and the growth of the Army's Future Combat System.[24] Further, DOD has reduced the number of C-5 *Galaxies* planned for upgrade with new engines and other enhancements that were expected to bolster capability of the C-5 fleet to levels required by MCS-2005. As a result, General Lichte stated that the current program of record for the Air Force's strategic airlift fleet of C-5s and C-17s falls short of the organic strategic airlift capability of 33.95 million ton miles day (MTM/D) requirement.[25]

The MCS called for the same level of CRAF contribution to total airlift capabilities (20.5 million ton miles per day of the overall 54.5 million ton miles per day objective) as required in the prior study.[26] However, DOD's projected use of CRAF to fulfill total airlift needs has increased from roughly 12 MTM/D in the late 1980s to roughly 20 MTM/D in 2005. Further, this increase in capacity has occurred gradually, and many view DOD's requirement for CRAF as being stable over this 19- year span. As *Figure 1* indicates, commercial aircraft committed to CRAF exceed DOD requirements. Thus, any foreseeable increases in CRAF requirements are unlikely to result in shortfalls of commercial aircraft committed to CRAF.

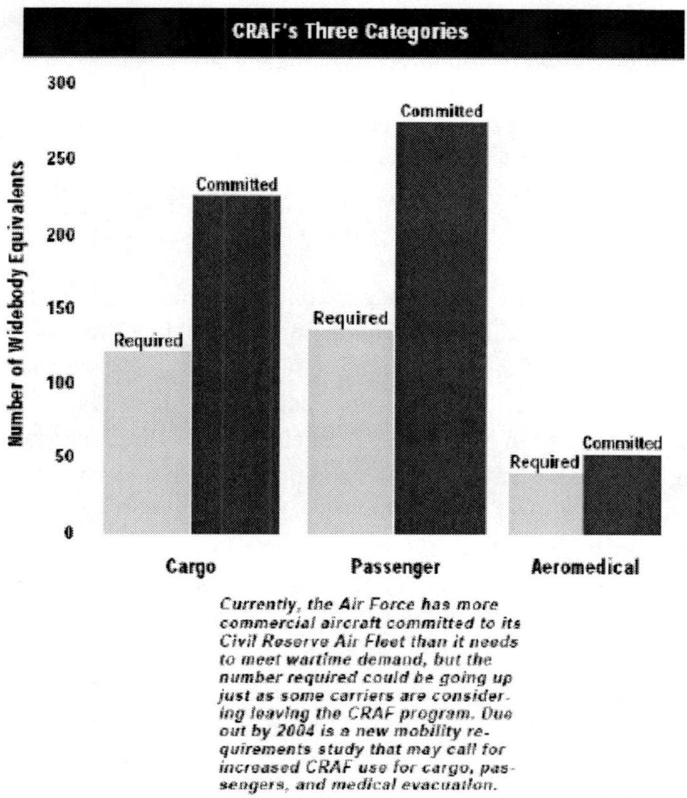

Source: Air Force Magazine, February 2003, p. 28.

Figure 1. CRAF Requirements and Commitments.

Potential C-17 Acquisition. Some favor acquisition of additional C-17s[27] to meet potential current and future strategic airlift requirement shortfalls. For example, in March 2008, General Norton A. Schwartz, Commander of U.S. Transportation Command, stated that based on the C-5 Reliability Enhancement and Re-engining Program being reduced and recertified by DOD, he believes DOD needs a fleet of 111 C-5s and 205 C-17s.[28] Further, the Air Force's FY2009 Unfunded Priority List contained a request for 15 additional C-17s.[29] In contrast, the Administration's FY2009 budget request did not contain funding for new C-17s, nor did it request funding to close the C-17 production line.[30] However, during congressional testimony, Gen. Schwartz cautioned that too large of an organic airlift fleet could potentially hurt the CRAF program in the future when he stated,

> One of the things that you hold me accountable for is sort of maintaining the balance between the organic fleet and the commercial capability. And as I mentioned in my opening remarks, I caution about overbuilding the organic fleet; because if that occurs, it can competes in peace time with that preference cargo, the incentives that we offer our commercial partners. And so that's one of the reasons that I believe 205 is the right number of C-17s.[31]

BC-17. Because Air Force budget limitations make additional large-scale procurement of C-17s difficult to fund, some have suggested the design of a commercial version of the C-17 aircraft (BC-17) that might become part of the CRAF fleet. However, is there sufficient market for these aircraft to be commercially viable? In May of 2007, Boeing's C-17 Program Manager, Dave Bowman, stated, "we have several customers with money that have given us requests for proposals."[32] Some industry studies suggest that a commercial market for up to 10 C-17s may exist for use in heavy industry, mining, or similar endeavors, while Boeing believes there is market potential of "upwards of 100 aircraft."[33] On the other hand, at present, there are no orders for a commercial variant of the C-17.

KC-X Acquisition. Acquisition decisions regarding KC-X, the Air Force's next generation tanker program, may also affect future DOD CRAF needs and use.34 Both competitors for the KC-X program, the Northrop Grumman KC-30 based on the Airbus 330-200 and the KC-767 based on Boeing's 767-200, could add airlift capability compared to the KC-135s they are envisioned to replace. Table 2 summarizes the airlift capabilities of the KC-135 and potential KC-X replacements.

Table 2. KC-135 and Potential KC-X Airlift Capabilities

	KC-135	KC-30	KC-767
Passengers	54	226	200
Pallets	6	32	19
Defensive Sysems	No	Planned	Planned

Source: USAF, Northrop Grumman and Boeing. KC-135 information derived from KC-135 Stratotanker Fact Sheet, http://www.af.mil/factsheets/factsheet.asp?id=110], and the Air Force Handbook 2007.

Like their commercial counterparts, potential KC-X tankers will have limited oversized cargo capability, but a significant capability to transport passengers and cargo pallets. Thus, as the Air Force's KC-X tanker fleet potentially grows, DOD's day-to-day need for commercial airlift that participants in the CRAF program provide could potentially be reduced. However, because most tankers could be needed to perform air refueling during a potential crisis, DOD would likely still rely on the CRAF program to meet surges in airlift demand.

On February 29, 2008, the Air Force awarded the KC-X contract to Northrop Grumman. The initial $12.1 billion KC-X contract provides for the purchase the first 68 KC-45s[35] of the anticipated 179 aircraft.[36] On March 11, 2008, Boeing protested the Air Force's decision to the Government Accountability Office (GAO).[37] GAO has 100 days to evaluate the protest.[38]

Industrial Base/Financial Assistance to Air Carriers

All major passenger and cargo carriers participate in CRAF.[39] This strong participation can be inferred to reflect broad support for CRAF. The program is voluntary, and it appears logical that if the airlines didn't find participation to be in their interest, they would not participate. Every indication suggests that U.S. air carriers value CRAF and want to participate. *Table 3* illustrates the growth in CRAF participation over the last 10 years.

Table 3. Recent CRAF Participation

	1998	2002	2005	2007	2008
Carriers	35	33	40	37	35
Aircraft	657	927	1,126	1,360	1,262

Source: Departments of Defense and Transportation.

Following the terrorist attacks of September 11, 2001, many U.S. commercial air carriers struggled because of a lack of business and other factors. Today, rising fuel prices continue to pose a threat to the commercial airline industry.[40] As economic and financial conditions for commercial air carriers potentially worsen, the benefit of CRAF for the commercial sector has been increasingly discussed. It may be that if economic conditions remain difficult, pressure may build on DOD to use more commercial airlift, not necessarily to satisfy DOD needs, but to support the private sector.[41] Also, some, including the DOT, have proposed changes to Federal Aviation Administration (FAA) regulations that might potentially lead to increased foreign investment in U.S. airlines, including those that participate in CRAF.[42] While some support the additional capital that foreign investment could bring to the airline industry, others oppose the concept of allowing foreign corporations to yield increased influence over a sector of the U.S. economy that makes a significant contribution to our nation's defense.

SUMMARY OF RECENT LEGISLATIVE ACTIVITY

This section provides a summary of recent legislation regarding the Department of Defense's (DOD's) Civil Reserve Air Fleet (CRAF) program.

FY2008

The FY2008 National Defense Authorization Act (NDAA) contained three provisions that affected the CRAF (P.L. 110-181). First, Section 356 called for a comprehensive and independent assessment of CRAF. This assessment is designed to examine current and long-range issues associated with CRAF and make specific recommendations for preserving and improving the program. The FY2008 NDAA required a report to be delivered to congressional defense committee no later than April 2008. An excerpt of Section 356 is provided at *Appendix A*.

Second, Section 378 of the FY2008 NDAA extended authorization of the Aviation Insurance Program (AIP) from March 30, 2008, to December 30, 2103. As part of the AIP, the FAA offers a non-premium insurance program to air carriers that participate in the CRAF. The Congressional Budget Office estimated "that extending the CRAF program through 2013 would have no significant budgetary impact."[43]

Third, Section 1046 called for a DOD study on the size and mix of the airlift force to specifically include how the CRAF could potentially affect DOD's airlift fleet requirements. This report is expected to be completed by January 2009. An excerpt of Section 1046 is provided at *Appendix B*. Many expect this study to inform force structure decisions regarding the optimal mix of DOD's organic air mobility fleets and the CRAF.

FY2007

In FY2007, the Senate version of the FY2007 NDAA contained a provision (sec. 1052) that would allow the Department of Defense to guarantee higher minimum levels of business to U.S. Civil Reserve Air Fleet carriers than are currently authorized by law. However, the provision was not adopted in the final legislation (P.L. 109-364).

FY2006

Section 131 of the FY2006 NDAA contained a provision that required DOD to conduct an analysis of inter-theater airlift capabilities to include the impact of the CRAF on DOD's inter-theater airlift force structure requirements (P.L. 109-163).

APPENDIX A. FY2008 NATIONAL DEFENSE AUTHORIZATION ACT (P.L. 110-181), SECTION 356

Section 356 of the Conference Report (H. Rept. 110-477, December 6, 2007) to H.R. 1585 stated the following:

SEC. 356. INDEPENDENT ASSESSMENT OF CIVIL RESERVE AIR FLEET VIABILITY

(a) Independent Assessment Required- The Secretary of Defense shall provide for an independent assessment of the viability of the Civil Reserve Air Fleet to be conducted by a federally-funded research and development center selected by the Secretary.

(b) Contents of Assessment- The assessment required by subsection (a) shall include each of the following:

1) An assessment of the Civil Reserve Air Fleet as of the date of the enactment of this Act, including an assessment of —

(A) the level of increased use of commercial assets to fulfill Department of Defense transportation requirements as a result of the increased global mobility requirements in response to the terrorist attacks of September 11, 2001;

(B) the extent of charter air carrier participation in fulfilling increased Department of Defense transportation requirements as a result of the increased global mobility requirements in response to the terrorist attacks of September 11, 2001;

(C) any policy of the Secretary of Defense to limit the percentage of income a single air carrier participating in the Civil Reserve Air Fleet may earn under contracts with the Secretary during any calendar year and the effects of such policy on the air carrier industry in peacetime and during periods during which the Armed Forces are deployed in support of a contingency operation for which the Civil Reserve Air Fleet is not activated; and

(D) any risks to the charter air carrier industry as a result of the expansion of the industry in response to contingency operations resulting in increased demand by the Department of Defense.

2) A strategic assessment of the viability of the Civil Reserve Air Fleet that compares such viability as of the date of the enactment of this Act with the projected viability of the Civil Reserve Air Fleet 5, 10, and 15 years after the date of the enactment of this Act, including for activations at each of stages 1, 2, and 3 —

(A) an examination of the requirements of the Department of Defense for the Civil Reserve Air Fleet for the support of operational and contingency plans, including any anticipated changes in the Department's organic airlift capacity, logistics concepts, and personnel and training requirements;

(B) an assessment of air carrier participation in the Civil Reserve Air Fleet; and

(C) a comparison between the requirements of the Department described in subparagraph (A) and air carrier participation described in subparagraph (B).

3) An examination of any perceived barriers to Civil Reserve Air Fleet viability, including —

(A) the operational planning system of the Civil Reserve Air Fleet;
(B) the reward system of the Civil Reserve Air Fleet;
(C) the long-term affordability of the Aviation War Risk Insurance Program;
(D) the effect on United States air carriers operating overseas routes during periods of Civil Reserve Air Fleet activation;
(E) increased foreign ownership of United States air carriers;
(F) increased operational costs during activation as a result of hazardous duty pay, routing delays, and inefficiencies in cargo handling by the Department of Defense;
(G) the effect of policy initiatives by the Secretary of Transportation to encourage international code sharing and alliances; and
(H) the effect of limitations imposed by the Secretary of Defense to limit commercial shipping options for certain routes and package sizes.

4) Recommendations for improving the Civil Reserve Air Fleet program, including an assessment of potential incentives for increasing participation in the Civil Reserve Air Fleet program, including establishing a minimum annual purchase amount during peacetime.

(c) Submission to Congress- Upon the completion of the assessment required under subsection (a) and by not later than April 1, 2008, the Secretary shall submit to the congressional defense committees a report on the assessment.
(d) Comptroller General Report- Not later than 90 days after the report is submitted under subsection (c), the Comptroller General shall conduct a review of the assessment required under subsection (a).

APPENDIX B. FY2008 NATIONAL DEFENSE AUTHORIZATION ACT (P.L. 110-181), SECTION 1046

Section 1046 of the Conference Report (H. Rept. 110-477, December 6, 2007) to H.R. 1585 stated the following:

SEC. 1046. STUDY ON SIZE AND MIX OF AIRLIFT FORCE.

(a) Study Required- The Secretary of Defense shall conduct a requirements-based study on alternatives for the proper size and mix of fixed-wing intratheater and intertheater airlift assets to meet the National Military Strategy for each of the following timeframes: fiscal year 2012, 2018, and 2024. The study shall —

1) focus on organic and commercially programmed airlift capabilities;
2) analyze the full-spectrum lifecycle costs of the various alternatives for organic models of each of the following aircraft: C-5A/B/C/M, C-17A, KC-X, KC-10, KC-135R, C-130E/H/J, Joint Cargo Aircraft; and
3) incorporate the augmentation capability, viability, and feasibility of the Civil Reserve Air Fleet during activation stages I, II, and III.

(b) Use of FFRDC- The Secretary shall select, to carry out the study required by subsection (a), a federally funded research and development center that has experience and expertise in conducting similar studies.

(c) Study Plan- The study required by subsection (a) shall be carried out under a study plan. The study plan shall be developed as follows:

1) The center selected under subsection (b) shall develop the study plan and shall, not later than 60 days after the date of enactment of this Act, submit the study plan to the congressional defense committees, the Secretary, and the Comptroller General of the United States.
2) The Comptroller General shall review the study plan to determine whether it is complete and objective, and whether it has any flaws or weaknesses in scope or methodology, and shall, not later than 30 days after receiving the study plan, submit to the Secretary and the center a report that contains the results of that review and provides any recommendations that the Comptroller General considers appropriate for improvements to the study plan.
3) The center shall modify the study plan to incorporate the recommendations under paragraph (2) and shall, not later than 45 days after receiving that report, submit to the Secretary and the congressional defense committees a report on those modifications. The report shall describe each modification and, if the modifications do not incorporate one or more of the recommendations, shall explain the reasons for not doing so.

(d) Elements of Study Plan- The study plan required by subsection (c) shall address, at minimum, the following:

1) A description of lift requirements and operating profiles for airlift aircraft required to meet the National Military Strategy, including assumptions regarding the following:
   (A) Current and future military combat and support missions.
   (B) The planned force structure growth of the military services.
   (C) Potential changes in lift requirements, including the deployment of the Future Combat Systems by the Army.
   (D) New capability in airlift to be provided by the KC(X) aircraft and the expected utilization of such capability, including its use in intratheater lift.
   (E) The utilization of intertheater lift aircraft in intratheater combat mission support roles.
   (F) The availability and application of Civil Reserve Air Fleet assets in future military scenarios.
   (G) Air mobility requirements associated with the Global Rebasing Initiative of the Department of Defense.
   (H) Air mobility requirements in support of worldwide peacekeeping and humanitarian missions.
   (I) Air mobility requirements in support of homeland defense and national emergencies.
   (J) The viability and capability of the Civil Reserve Air Fleet to augment organic forces in both friendly and hostile environments.
   (K) An assessment of the Civil Reserve Air Fleet to adequately augment the organic fleet as it relates to commercial inventory management restructuring in response to future commercial markets, streamlining of operations, efficiency measures, or downsizing of the participant.
2) An evaluation of the state of the current airlift fleet of the Air Force, including assessments of the following:
   (A) The extent to which the increased use of airlift aircraft in on-going operations is affecting the programmed service life of the aircraft of that fleet.
   (B) The adequacy of the current airlift force, including whether or not a minimum of 299 strategic airlift aircraft for the Air Force is sufficient to support future expeditionary combat and non-combat missions, as well as domestic and training mission demands consistent with the requirements of meeting the National Military Strategy.

(C) The optimal mix of C-5 and C-17 aircraft for the strategic airlift fleet of the Air Force, to include the following:

i. The cost-effectiveness of modernizing various iterations of the C-5A and C-5B/C aircraft fleet versus procuring additional C-17 aircraft.

ii. The military capability, operational availability, usefulness, and service life of the C-5A/B/C/M aircraft and the C-17 aircraft. Such an assessment shall examine appropriate metrics, such as aircraft availability rates, departure rates, and mission capable rates, in each of the following cases:

I. Completion of the Avionics Modernization Program and the Reliability Enhancement and Reengining Program.

II. Partial completion of the Avionics Modernization Program and the Reliability Enhancement and Reengining Program, with partial completion of either such program being considered the point at which the continued execution of each program is no longer supported by the cost-effectiveness analysis.

iii. At what specific fleet inventory for each organic aircraft, to include air refueling aircraft used in the airlift role, would it impede the ability of Civil Reserve Air Fleet participants to remain a viable augmentation option.

(D) An analysis and assessment of the lessons that may be learned from the experience of the Air Force in restarting the production line for the C-5 aircraft after having closed the line for several years, and recommendations for the actions that the Department of Defense should take to ensure that the production line for the C-17 aircraft could be restarted if necessary, including —

i. an analysis of the methods that were used and costs that were incurred in closing and re-opening the production line for the C-5 aircraft;

ii. an assessment of the methods and actions that should be employed and the expected costs and risks of closing and re-opening the production line for the C-17 aircraft in view of that experience.

Such analysis and assessment should deal with issues such as production work force, production facilities, tooling, industrial base suppliers, contractor logistics support versus organic maintenance, and diminished manufacturing sources.

(E) Assessing the military capability, operational availability, usefulness, service life and optimal mix of intra-theater airlift aircraft, to include —

i. the cost-effectiveness of procuring the Joint Cargo Aircraft versus procuring additional C-130J or refurbishing C-130E/H platforms to meet intra-theater airlift requirements of the combatant commander and component commands; and

ii. the cost-effectiveness of procuring additional C-17 aircraft versus procuring additional C-130J platforms or refurbishing C-130E/H platforms to meet intra-theater airlift requirements of the combatant commander and component commands.

3) Each analysis required by paragraph (2) shall include —

(A) a description of the assumptions and sensitivity analysis utilized in the study regarding aircraft performances and cargo loading factors; and

(B) a comprehensive statement of the data and assumptions utilized in making the program life cycle cost estimates and a comparison of cost and risk associated with the optimally mixed fleet of airlift aircraft versus the program of record airlift aircraft fleet.

(E) Utilization of Other Studies- The study required by subsection (a) shall build upon the results of the 2005 Mobility Capabilities Studies, the on-going Intra-theater Airlift Fleet Mix Analysis, the Intra-theater Lift Capabilities Study, the Joint Future Theater Airlift Capabilities Analysis, and other appropriate studies and analyses, such as Fleet Viability Board Reports or special aircraft assessments. The study shall also include any testing data collected on modernization, recapitalization, and upgrade efforts of current organic aircraft.

(F) Collaboration With United States Transportation Command- In conducting the study required by subsection (a) and preparing the report required by subsection (c)(3), the center shall collaborate with the commander of the United States Transportation Command.

(G) Collaboration With Cost Analysis Improvement Group- In conducting the study required by subsection (a) and constructing the analysis required by subsection (a)(2), the center shall collaborate with the Cost Analysis Improvement Group of the Department of Defense.

(H) Report- Not later than January 10, 2009, the center selected under subsection (b) shall submit to the Secretary and the congressional defense committees a report on the study required by subsection (a). The report shall be submitted in unclassified form, but shall include a classified annex.

End Notes

[1] Executive Order No 10219, Federal Register 16, "Defining Certain Responsibilities of Federal Agencies With Respect to Transportation and Storage." March 2, 1951, p. 1004- 1005.

[2] Mary E. Chenoweth, "The Civil Reserve Air Fleet and Operation Desert Shield/Desert Storm: Issues for the Future," RAND, Santa Monica, CA, 1993. pp. 4-5.

[3] "Outsized cargo" exceeds the dimensions of oversized cargo and requires the use of a C-5 or C-17 aircraft or surface transportation. "Oversized cargo" exceeds the dimensions of a standard (463L) pallet, but is air transportable on the C-5, C-17, C-130, KC-10, and most civilian contract cargo carriers. (Joint Publication 1-02, DOD Dictionary of Military and Associated Terms, amended through October 17, 2007, pp. 401-402).

[4] Civil Reserve Air Fleet (Fact Sheet), U.S. Air Force, July 2007, at [http://www.af.mil/factsheets/factsheet.asp?id=173]

[5] Ibid.

[6] Ibid.

[7] Civil Reserve Air Fleet (Fact Sheet), U.S. Air Force, July 2007.

[8] Ibid.

[9] "Air Force announces CRAF contracts valued at $2.2 billion." (Release Number: 570905) Air Force Print News Today. September 13, 2005.

[10] Military Readiness. Civil Reserve Air Fleet Can Respond as Planned, but Incentives May Need Revamping. General Accounting Office (GAO-03-278) December 2002. p.15.

[11] Civil Reserve Air Fleet (Fact Sheet), U.S. Air Force, July 2007.

[12] Ibid.

[13] Chuck Roberts, Master Sergeant, USAF, "Civil Reserve Air Fleet Stage I Activates for Only the Second Time in History," Airman, April 2003, at [http://www.af.mil/news/ airman/0403/world7.html].

[14] Michael Lowder, Letter to CRAF Program Participants, Enclosure 1, March 11, 2008, at [http://www.dot.gov/ost/oet/craf/CR0308.xls]

[15] GAO/NSIAD-96-125, "Military Airlift: Observations on the Civil Reserve Air Fleet Program," March 1996, p. 2, at [http://www.gao.gov/archive/1996/ns96125.pdf].

[16] Robert C. Owen, Professor, Embry-Riddle Aeronautical University-Daytona Beach, FL, "Transport Trade-offs," written in response to a previous letter to the editor in "Correspondence," Aviation Week and Space Technology, October 8, 2007, p. 8.

[17] Ibid., p. 7.

[18] Jean Gebman, et al. Finding the Right Mix of Military and Civil Airlift, Issues and Implications. Volume 1. Executive Summary. RAND. Santa Monica, CA.1994. p. 21.

[19] Micheline Maynard. "Airlines Moving More Troops This Month." New York Times. January 25, 2005.

[20] RAND op cit. p. 22.

[21] Jason Sherman. "CENTCOM Taps Commercial Air Carriers to Cut Iraq Cargo Costs." Inside the Air Force. July 21, 2006.

[22] See for example, Cynthia Di Pasquale. "Pentagon Studies CRAF Sustainability and Incentives to Industry."Inside the Air Force. October 8, 2004.

[23] See CRS Report RL34264, Strategic Airlift Modernization: Analysis of C-5 Modernization and C-17 Acquisition Issues, by William Knight and Christopher Bolkcom for more information.

[24] Gen. Arthur L. Lichte (USAF), Written Statement, Hearing on United States Transportation Command Posture and Air Force Mobility Aircraft Programs, Subcommittee on Air and Land Forces, House Armed Services Committee, April 1, 2008, p. 8.

[25] Ibid, p.4.

[26] Unlike previous mobility studies, DOD's Mobility Capabilities Study 2005 did not provide a specific quantitative estimate of airlift requirements. Some have asserted that this omission reduces the value of the MCS and have called for another, requirements-driven study.

[27] See CRS Report RS22763, Military Airlift: C-17 Program Background, by William Knight and Christopher Bolkcom for more information about the C-17.

[28] Transcript from the Hearing of the Senate Armed Services Subcommittee on Seapower on Fiscal Year 2009 Budget for the Department of Defense Strategic Lift Programs, March 12, 2008 obtained through [http://www.cq.com].

[29] "FY2009 Unfunded Requirements List," U.S. Air Force, SAF/FMB, Budget and Appropriations Liaison Office, February 2008, p. 2, at [http://dailyreport.afa.org/NR/rdonlyres/BAD928B9-0927-4628-BE43-14992CB464C1/0/FY09URL.pdf].

[30] Marc V. Schanz, "The C-17 and the Airlift Question," Daily Report, Air Force Magazine, February 6, 2008, at [http://dailyreport.afa.org/AFA/Features/modernization/ box020608 airlift.htm].

[31] Transcript from the Hearing of the Senate Armed Services Subcommittee on Seapower on Fiscal Year 2009 Budget for the Department of Defense Strategic Lift Programs, March 12, 2008 obtained through [http://www.cq.com].

[32] Guy Norris, online at Flightglobal.com, May 23, 2007.

[33] Ibid.

[34] See CRS Report RL34398, Air Force Air Refueling: The KC-X Acquisition Program, by William Knight and Christopher Bolkcom.

[35] The Air Force plans to designate its KC-X tanker as the KC-45A.

[36] "Tanker Contract Award Announced," Air Force Print News, February 29, 2008, online at [http://www.af.mil/news/story.asp?id=123088392].

[37] Boeing News Release, "Boeing Protests U.S. Air Force Tanker Contract Award," March 11, 2008, at [http://www.boeing.com/ids/globaltanker/news/2008/q1/ 080311b_nr.html].

[38] GAO Press Statement, "GAO Statement Regarding Protest Filed by Boeing Company with the Government Accountability Office," March 11, 2008, at [http://www.gao.gov/press/pressstmt-boeing0311.pdf].

[39] United, American, Delta, Northwest, US Airways, Southwest, UPS, and FEDEX respectively.

[40] Graham Bowley, "Frontier Airlines Files for Bankruptcy," The New York Times, April 12, 2008, at [http://www.nytimes.com/2008/04/12/business/11cnd-frontier.html? ex=1208577600&en=f41b5269883e5591&ei=5123&partner=BREITBART].

[41] See for example Scott McCartney. "Flying Troops for the Pentagon Gives Airlines a Financial Boost." Wall Street Journal. May 14, 2003. Micheline Maynard. "Pentagon's War Needs Are a Lifeline for Airlines." New York Times. December 26, 2003. Micheline Maynard. "Pentagon's Needs Offer Airlines a Lift." New York Times. January 22, 2005.

[42] See for example 70 Fed. Reg. 67,389 (November 7, 2005), a Notice of Proposed Rulemaking (NPRM) seeking comments on a proposal to clarify the policy of requiring "actual control" of a U.S. air carrier to be by a citizen of the United States.

[43] Congressional Budget Office Cost Estimate, "H.R. 4986: National Defense Authorization Act for Fiscal Year 2008," January 25, 2008, p. 12.

In: Civil Reserve Air Fleet
Editor: Soren M. Jonsson

ISBN: 978-1-62808-782-6

Chapter 2

ISSUES REGARDING THE CURRENT AND FUTURE USE OF THE CIVIL RESERVE AIR FLEET*

David Arthur

OVERVIEW

To support military operations in Afghanistan and Iraq, the Department of Defense (DoD) has made substantial use of commercial air transportation provided by carriers that participate in the Civil Reserve Air Fleet (CRAF) program. Over the 2002–2006 period, DoD's total expenditures for airlift services provided by CRAF carriers averaged about $2.1 billion annually, a nearly fourfold increase over the average during the previous five years. Carriers that participate in the CRAF program are eligible to receive that business from DoD in exchange for making their aircraft and air crews available on short notice to support rapid, large-scale deployments of military forces. DoD is anticipating a substantial decrease in the need for commercial airlift services when operations in the Middle East ultimately wind down. To maintain carriers' incentives to participate in the CRAF program, DoD submitted a proposal as part of its fiscal year 2008 budget request that would allow it to guarantee CRAF carriers more business at the beginning of a fiscal year than it is currently authorized to guarantee.

* David Arthur prepared this report which was released by the Congressional Budget Office, October 2007.

At the request of the Subcommittee on Readiness of the House Armed Services Committee, the Congressional Budget Office (CBO) examined how the increased use of commercial airlift services over the past few years and the potential increase in the guaranteed portion of DoD's airlift business might affect the U.S. air transportation industry and the military's ability to attract participants to the CRAF program over the long term. CBO found that although DoD's current use of commercial airlift services is at a historically high level, that use is still small in relation to the overall size of the nation's air transportation industry. For example, U.S. air carriers' revenues from DoD's cargo business in 2005 were less than 5 percent of carriers' total cargo revenues in that year, and their revenues from DoD's passenger business were less than 2 percent of total passenger revenues. Moreover, those revenues were not spread evenly across the industry but instead were concentrated among a few small airlines. Consequently, a decrease in DoD's use of commercial airlift services might not have a large adverse effect on the air transportation industry as a whole. However, it could result in a significant loss of revenues for air carriers that currently rely on DoD for a large share of their business if those carriers could not recoup that loss by serving more commercial customers.

During peacetime, the business that DoD offers to CRAF carriers is divided into a "fixed buy," based on DoD's known needs for airlift services during the coming year, and an "expansion buy," to cover additional requirements as they arise. The fixed buy is a particularly strong incentive for carriers to participate in the CRAF program because it is set at a guaranteed level at the beginning of each fiscal year.[1] If DoD's projections of its needs for commercial airlift support after the end of operations in Iraq and Afghanistan are accurate, the fixed-buy por- tion of the CRAF program—and therefore the incentive for businesses to participate in the program under current policies—will be considerably smaller than it was before the start of the current conflicts in the Middle East.[2]

To maintain the pre-2002 levels of guaranteed business (and thus the incentive), DoD has proposed setting those guaranteed amounts for CRAF participants at no more than 80 percent of its average total commercial business—that is, fixed and expansion buys—over the preceding five years (omitting years of unusually high demand). The proposal would allow DoD initially to provide guarantees that are up to 50 percent higher than those provided before 2002, CBO estimates. Because the proposal would allow guaranteed payments to be based on expected rather than known requirements, the government would run some risk of having to pay for services that it might not use.

The amount of guaranteed DoD business before 2002 proved to be a sufficient incentive to attract CRAF participants—in fact, the number of aircraft committed to the program was larger than the number required under DoD's wartime deployment scenarios. Nevertheless, whether those levels will be enough of an incentive in the future is unclear. On the one hand, industry analysts predict substantial growth in air transportation revenues in the coming decade—which would make DoD's peacetime business smaller relative to the overall market and therefore less worth the risk of the disruption to carriers' commercial operations that results when DoD calls up aircraft for a large-scale CRAF activation.[3] On the other hand, because a larger number of aircraft would be operated by a larger air transportation industry, the temporary loss of aircraft to an activation under the CRAF program might be less disruptive to commercial operations than it would have been in the past.

The Civil Reserve Air Fleet Program

The CRAF program was created to augment DoD's fleet of military transport aircraft during times of high demand for airlift services. Under the CRAF business model, U.S. commercial air carriers contractually commit aircraft and air crews for use by DoD during times of crisis. In exchange for that commitment, DoD makes peacetime business available to carriers that participate in the program.

The Structure of the Civil Reserve Air Fleet

The fleet is divided into three main segments—international, national, and aeromedical evacuation—that primarily reflect the type of aircraft required for operations. Aircraft in each of the segments can be activated in three stages, depending on the airlift services that are needed to augment the military's transport capacity.

CRAF Segments

Most of the aircraft in the Civil Reserve Air Fleet are committed to the international segment, which is divided into a long-range section (for aircraft that have a minimum range of 3,500 nautical miles) and a short-range section (for aircraft that have less endurance; see Table 1). Most of the aircraft in the CRAF's long-range section are passenger planes because the Air Force's long-range transports—the C-17s and C-5s—are configured to be more useful for carrying cargo.

The much smaller national segment of the fleet also has two sections: a domestic section for most transportation within the United States and a small section for the support of operations in Alaska.[4] The domestic section includes only passenger aircraft, and the Alaskan section, only cargo aircraft. The aeromedical evacuation segment includes Boeing 767 aircraft that can be rapidly reconfigured into flying ambulances.

Stages of CRAF Activation

The commander of the United States Transportation Command (USTRANSCOM), with the concurrence of the Secretary of Defense, has the authority to activate the Civil Reserve Air Fleet, which can be called up incrementally in three stages.

- Stage I covers minor operations or operations in which adequate time is available so that a small augmentation of the military's fleet is sufficient to move the required people or cargo. A Stage I CRAF activation of long-range international cargo and passenger aircraft occurred from August 1990 to January 1991 in support of Operation Desert Shield, and a Stage I activation of long-range international passenger aircraft occurred from February to June 2003 in support of Operation Iraqi Freedom.
- Stage II is tailored for a major theater war that requires rapid deployment of forces. In January 1991, the long-range international segment was activated to Stage II for both passenger and cargo aircraft in support of Operation Desert Shield/Desert Storm.
- Stage III would be required if the military had to fight more than one major theater war at the same time or operate in a larger crisis. A Stage III CRAF activation has never occurred. It was seriously considered after the Desert Storm air war began, in late January 1991, but was rendered unnecessary by the short duration of the conflict.

The Stage I and II examples given above are the only times that the Civil Reserve Air Fleet has been activated. As of June 2007, a total of 73 aircraft were committed to Stage I; 304 to Stage II; and 1,363 to Stage III.

Table 1. Aircraft Committed to the Civil Reserve Air Fleet in June 2007, by Segment

		Cargo Aircraft			*Passenger Aircraft*	
	Stage I	*Stage II*	*Stage III*	*Stage I*	*Stage II*	*Stage III*
International Segment						
Long-range section	31	79	300	42	147	690
Short-range section	n.a.	12	12	n.a.	13	270
National Segment						
Domestic services section	n.a.	0	0	n.a.	24	37
Alaskan section	n.a.	4	4	n.a.	0	0
Aeromedical Evacuation Segment	n.a.	0	0	n.a.	25	50

Source: Congressional Budget Office using information from the United States Transportation Command.

Notes: Internal divisions of the Civil Reserve Air Fleet (segments and sections) reflect the unique capabilities required of committed aircraft. n.a.

Contractual Relationship Between Carriers and the Government

Under the CRAF program, U.S. air carriers commit aircraft and crews to the specific segments described above. When DoD activates the Civil Reserve Air Fleet, the carriers continue to operate the aircraft, but their scheduling comes under the control of the Air Force's Air Mobility Command. Although the government pays carriers a negotiated rate per ton-mile or passenger-mile for the use of the aircraft during an activation, a carrier's commercial business may be significantly disrupted by the temporary unavailability of its aircraft.

For example, after the Stage II CRAF activation during Operation Desert Shield/Desert Storm, executives at airlines that participated in the CRAF program expressed concern about the long-term loss, to competitors, of the business that they had been forced to forgo because their aircraft were committed to the military's operations.5 Such concerns were significant enough to result in the CRAF program's not meeting its goals for participation during the two years following Desert Shield/Desert Storm. The government responded by strengthening the incentives for participation, including guaranteeing a minimum level of use (and thus a minimum payment) for activated aircraft and changing rate structures in the program to account for unexpected operations costs, such as those experienced during Desert

Shield/Desert Storm. The primary incentive for a carrier to participate despite the risk of a disruptive call-up remained the opportunity to receive the government's peacetime business in proportion to the airlift capacity that the carrier had committed to the CRAF program. (The term "peacetime business" here denotes airlift services purchased by the government when the Civil Reserve Air Fleet is not formally activated.

Peacetime business may, nevertheless, support combat operations.)

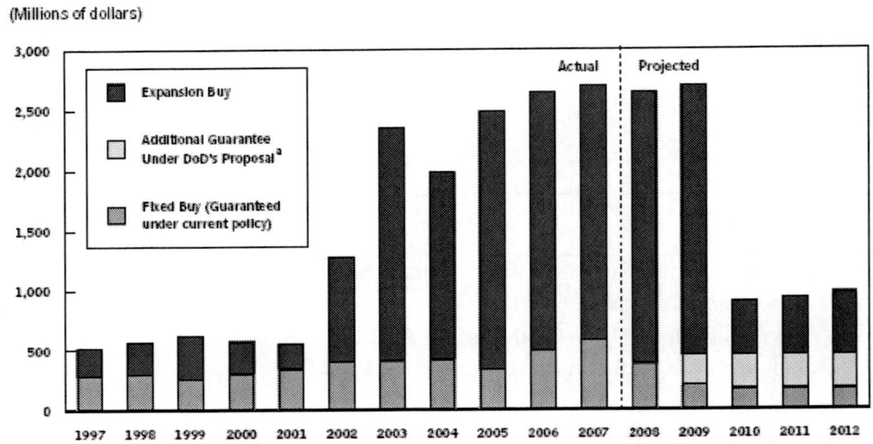

Source: Congressional Budget Office based on historical and projected data on fixed and total buys from the United States Transportation Command.

Note: CRAF carriers are offered a "fixed buy" based on known airlift needs for the coming year and an "expansion buy" for additional needs as they arise. Under current policy, the fixed buy is guaranteed at the beginning of each fiscal year.

a. CBO estimates.

Figure 1. Actual and Projected Expenditures for the Civil Reserve Air Fleet.

As noted, the government's peacetime business is divided into a fixed buy and an expansion buy. The fixed buy covers airlift support that can be specifically identified for the coming year. For example, a base in Germany might require a known number of transport flights each week to carry out its routine operations. The expansion buy covers other airlift needs that may arise, especially support for contingency operations, for which specific requirements are not known in advance. The distinction between the fixed buy and the expansion buy is important because the government guarantees payments to CRAF program participants in the amount of the fixed buy at the beginning of each fiscal year. Those guaranteed payments are a particularly attractive incentive to carriers to participate in the Civil Reserve Air Fleet because they can count on those funds in formulating their annual business plans. However, during the year, carriers are paid only when services are rendered; payment for the unused portion of the guarantee, if any, would be disbursed at the end of the year.

Implications of DoD's Recently Increased Use of CRAF Carriers

Since the beginning of the U.S. military's operations in Iraq and Afghanistan, DoD has been using more commercial airlift services than it did in previous years (see Figure 1). Although the Civil Reserve Air Fleet was activated only briefly, from February to June 2003, the government's annual expenditures for commercial airlift services provided through the CRAF program from 2002 through today have been nearly four times greater than the annual average during the 1997–2001 period. Because the CRAF program was designed to give DoD a "surge" airlift capacity for short periods, the military's prolonged use of commercial aircraft over the past several years to augment its airlift capacity has raised concerns about whether the commercial industry is becoming too reliant on DoD's business, a dependence that could leave it economically vulnerable when DoD's airlift needs eventually return to levels more typical of peacetime. However, CBO's analysis indicates that the U.S. commercial air transportation industry as a whole is sufficiently large that a decrease in business from the government might not have significant adverse implications. Nevertheless, some small cargo and passenger CRAF participants could find themselves in trouble because they currently rely on DoD for a large portion of their revenues.[6]

Cargo Carriers

DoD's use of the Civil Reserve Air Fleet during the past several years has accounted for an average of less than 5 percent of the Air Transport Association's estimate of total cargo revenues of U.S. air carriers.7 Although that percentage is about three times larger than the approximately 1.5 percent share of revenues seen prior to 2002, the share of total cargo revenues attributable to current CRAF participants has remained small relative to the revenues of the entire air cargo industry.

As Figure 2 shows, the largest and fastest-growing portion of air cargo revenues comes from transporting express and other time-definite freight, a type of cargo usually handled by door-to-door carriers (for example, United

Parcel Service and FedEx). Although such carriers have made significant commitments to the CRAF program, they most often opt to resell the DoD business for which they are eligible, directing it to charter cargo carriers. In 2006, for example, commitments by the large express and time-definite freight carriers represented between 35 percent and 55 percent of the cargo aircraft available in Stages I to III of CRAF's long-range international segment. Those carriers, however, transported less than 4 percent (as measured by revenues earned) of the cargo moved for DoD that year. Most of DoD's peacetime business goes to charter air cargo carriers, which form a smaller segment of the air cargo industry. Since the onset of operations in Afghanistan, in 2001, DoD's cargo business has averaged about 30 percent of the revenues of U.S. charter air cargo firms, up from about 10 percent during the preceding five years.

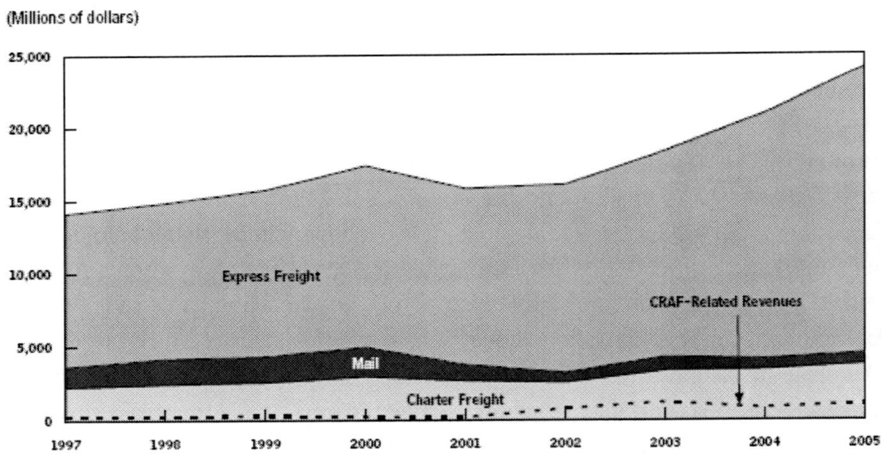

Source: Congressional Budget Office based on information from the Air Transport Association's 2006 Economic Report and the United States Transportation Command.

Note: CRAF = Civil Reserve Air Fleet.

Figure 2. Revenues of U.S. Cargo Air Carriers.

Although cargo carriers have clearly seen a sizable increase recently in the business they receive from the government, there is no evidence that carriers have added aircraft to their fleets that they could not sustain if the government's business were to diminish. For example, the three companies that received about 80 percent of DoD's cargo airlift expenditures for 2006—Atlas Air/Polar Air, Evergreen International, and ASTAR—do not appear to have significantly enlarged their fleets over the past several years (see Figure

3). That stability suggests that even those companies could accommodate a reduction in government-provided business.

Passenger Carriers

In the case of passenger carriers, the greater use of the Civil Reserve Air Fleet over the past several years has accounted for less than 2 percent of total air passenger revenues, on average—a smaller share than that observed for cargo carriers (see Table 2). As is the case with cargo carriers, though, that business has been concentrated among a few smaller carriers; in 2006, for example, four such carriers accounted for more than 80 percent of CRAF passenger revenues.

Historically, DoD's business has been a bigger fraction of U.S. charter passenger revenues than of charter cargo revenues. For the period 1997 to 2001, the use of CRAF funds to transport passengers averaged slightly more than 20 percent of the total value of passenger charters; since 2002, that average has risen to more than 50 percent. The increase stems from the large number of troops being rotated to and from Afghanistan and Iraq as well as from DoD's preference for carrying cargo on military aircraft. As a result, charter passenger carriers would potentially be more vulnerable to the effects of a drop in DoD's business.

Table 2. Estimated Revenues of U.S. Passenger Air Carriers

(Millions of dollars)	Industry wide Services		
Year	Scheduled	Charter	CRAF Percentage
1997	79,540	1,500	0.3
1998	81,052	1,600	0.4
1999	84,383	1,700	0.4
2000	93,622	1,950	0.4
2001	80,947	1,800	0.4
2002	73,577	1,700	0.7
2003	77,379	2,250	1.5
2004	85,646	2,250	1.3
2005	93,500	2,430	1.5
2006	101,208	2,200	1.4

Source: Congressional Budget Office. Notes: Data on scheduled services were drawn from the Air Transport Association's (ATA's) 2007 Economic Report. CBO estimated revenues for charter services on the basis of ATA data; it calculated the percentages of revenues attributable to the use of CRAF-related funds on the basis of data from the United States Transportation Command.

CRAF = Civil Reserve Air Fleet.

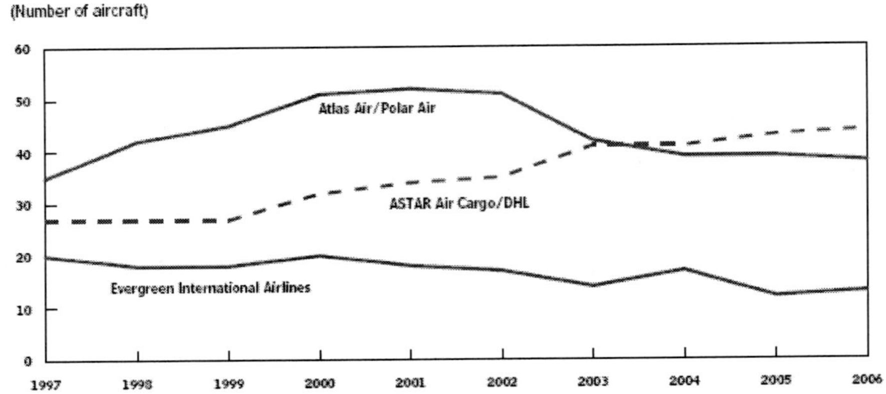

Source: Congressional Budget Office based on information from the Air Transport Association's annual Economic Reports.

Figure 3. Fleet Sizes of Selected Cargo Air Carriers.

Reliance of Individual Carriers on Business from DoD

In its use of the CRAF program, DoD tries to ensure that individual participants do not rely too heavily on the government's peacetime business because a carrier that dedicated a large portion of its capacity to DoD during peacetime would have little additional capacity to contribute to a wartime surge. To address that issue, DoD has maintained a long-standing policy that no more than 40 percent of a CRAF carrier's revenues should come from the government. Carriers that exceed the 40 percent threshold may be penalized by lower limits on the amount of DoD's business that the department assigns to them in subsequent years. However, DoD may also grant waivers to that so-called 60/40 policy under certain circumstances. For example, if a carrier accepts additional business from the government to help DoD avoid a Stage I activation, DoD may exempt the payments for that business from the 60/40 calculation. In addition to ensuring that the military maintains airlift capacity for a surge, the 60/40 policy reduces the likelihood that individual carriers will become too reliant on business from DoD.

Of the 33 cargo and passenger air carriers that elected to participate in the CRAF program during fiscal year 2007, four fell below the threshold of at least 60 percent in commercial revenues (see Figure 4). The four below-threshold carriers operated only 52 aircraft, a small fraction of the more than 4,000 aircraft operated by all carriers participating in the CRAF program. Three of those four provided primarily passenger transportation to DoD. Not all small carriers participating in the CRAF program are as heavily reliant on

DoD's business as those four are. (For example, in terms of total revenues, the ten smallest carriers were all above the 60 percent threshold for commercial revenues.)

IMPLICATIONS OF CHANGING THE WAY THAT CRAF GUARANTEES ARE ESTABLISHED

DoD has requested the authority to change the method it uses to calculate the annual guaranteed level of business for CRAF carriers.[8] Currently, the guarantee is equal to the fixed-buy portion of the CRAF program, which is based on firmly known, well-established needs (such as routine missions to move people and cargo between overseas bases and the United States). As a result, there is little chance that the guarantee will exceed the military's actual requirements. Under DoD's proposal, the new guaranteed level of business would be "based on forecast needs but may not be for more than eighty percent of the annual average expenditures of airlift for the prior 5-year period. Unusually high demand years, such as during a conflict, will normally be omitted to obtain a more accurate forecast." In other words, under the proposal, the guaranteed level might include funds that would previously have been disbursed as part of an expansion buy.

From 1997 to 2001, before operations began in Afghanistan and Iraq, DoD's guaranteed fixed buy averaged about $300 million per year, and the expansion buy averaged about $280 million. DoD, however, expects that the volume of airlift services that would qualify as part of the fixed buy will diminish, primarily because the United States plans to reduce the number of military personnel stationed overseas. Indeed, the department's prediction of a 40 percent drop in fixed-buy levels is consistent with its plans to reduce the number of people stationed overseas by about 70,000, or 35 percent. Thus, under current policies, the guaranteed amount of CRAF funding could drop to about $180 million per year. (However, DoD expects that the decrease in the fixed buy will be partially offset by an increase in the expansion buy, because in responding to unanticipated crises, the military will probably have to dispatch troops from the United States instead of from foreign bases that are closer to likely areas of operations.)

In documents in support of its proposal, DoD indicates that the changes are necessary to maintain the pre-2002 level of guaranteed CRAF funding and retain an adequate incentive for carriers to participate in the program after the end of major operations in the Middle East. The guarantee limit that DoD proposes would initially allow significantly higher guarantees, however—80 percent of the average total CRAF buy for 1997 through 2001 would be nearly $450 million, which is about 50 percent greater than the $300 million average annual guarantee over the earlier period but less than total CRAF spending in any year during that period. (The wartime span of 2002 through 2008 would be excluded from the calculation.)

Because the need for peacetime commercial airlift support has changed only slowly over time, the Air Force's planners are likely to establish a conservative CRAF business guarantee to avoid committing funds unnecessarily to airlift services. For example, briefings from USTRANSCOM show that projections of annual guarantees beginning in 2009 are closer to the 1997–2001 average level of about $300 million, which is significantly lower than the maximum allowed under the proposed legislation. Yet whether even the higher guarantee levels would be adequate in the future to attract sufficient numbers of air carriers to the CRAF program is uncertain.

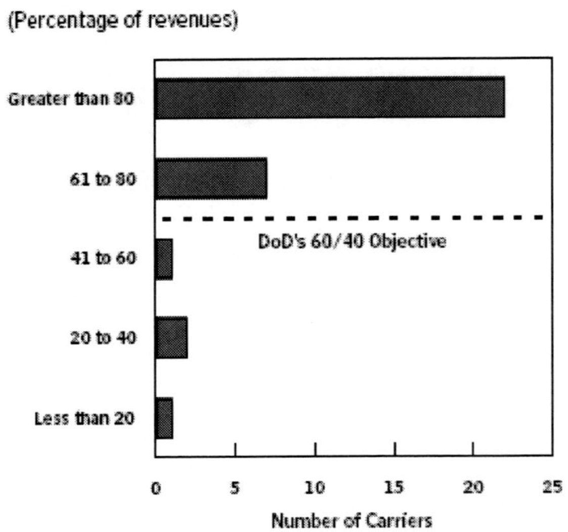

Source: Congressional Budget Office based on 2007 data from the United States Transportation Command.

Notes: The 60/40 objective refers to the Department of Defense's policy that no more than 40 percent of a CRAF carrier's revenues should come from the federal government.

CRAF = Civil Reserve Air Fleet.

Figure 4. Fraction of CRAF Carriers' Revenues from Commercial Customers.

The substantial growth predicted for the air transportation industry in coming years means that today's guarantee levels and perhaps even DoD's current total purchases of commercial airlift services will be a steadily diminishing piece of total U.S. air transportation revenues. As a result, air carriers may decide that the revenues available from DoD's CRAF peacetime business are not worth the risk of disruption in their commercial operations if the reserve fleet is activated and carriers' aircraft are diverted to support military operations.

Conversely, though, if the airlift capacity needed for the reserve fleet's various stages of operations does not increase, the aircraft that might be called up during a CRAF activation would be a significantly smaller fraction of the commercial fleet. Boeing, for example, predicts a doubling of the world's air freighter fleet over the next 20 years.[9] Air carriers might be more willing to participate in the Civil Reserve Air Fleet despite lower or stagnant guaranteed and total business levels if they determined that a CRAF activation posed a smaller risk of disruption. If peacetime business opportunities become an insufficient incentive to encourage carriers to participate in CRAF, DoD may be forced to find other means of obtaining a surge airlift capacity.

End Notes

[1] From 1997 through 2001, the fixed buy averaged slightly more than half of DoD's total CRAF business. Since the beginning of the wars in Afghanistan and Iraq, the fixed buy has averaged about 20 percent of all CRAF business.

[2] DoD expects smaller fixed buys in part because it plans to reduce the number of military personnel stationed overseas by more than a third—or about 70,000 people—over the next several years. Much of the fixed buy pays for routine missions to transport people and cargo between overseas bases and the United States. However, DoD anticipates that expansion buys will increase because it will be more costly to deploy forces from the United States than from overseas bases that would be nearer likely areas of operations.

[3] See, for example, Boeing's "World Air Cargo Forecast 2006– 2007," available at www.boeing.com/commercial/cargo/ index.html.

[4] As of June 2007, aircraft committed to the Alaskan section included propeller-powered DC-6 and L-100 aircraft suitable for operating from the kind of rugged, semiprepared airstrips often found in the Arctic. (The L-100 is a commercial version of the military's C-130 Hercules, and the DC-6 is a civilian version of the C-118 Liftmaster.) In contrast, the aircraft that carriers have committed to the other parts of the fleet are more-modern commercial jets.

[5] See James K. Matthews and Cora J. Holt, So Many, So Much, So Far, So Fast: United States Transportation Command and Strategic Deployment for Operation Desert Shield/Desert Storm (Department of Defense, Office of the Chairman of the Joint Chiefs of Staff and the United States Transportation Command, 1995).

[6] CBO's conclusions are similar to those of a large study by the Institute for Defense Analyses (IDA) that was completed in 2003. The study was conducted during the slump that occurred in the airline industry following the September 11, 2001, terrorist attacks. Nevertheless, the IDA study, which included extensive interviews with airline industry officials, concluded that DoD's commercial airlift business had "an insignificant impact on the financial health of the nation's largest air carriers." The study noted, however, that in the case of several smaller carriers, DoD's business "contributes to the carrier's long-term survival." See Institute for Defense Analyses, Sustaining the Civil Reserve Air Fleet (CRAF) Program (Alexandria, Va.: Institute for Defense Analyses, May 2003).

[7] Air Transport Association, Economic Report (Washington, D.C.: Air Transport Association, various years), available at www. airlines.org/products/pubs/product-detail.htm?Product=61.

[8] The current method of establishing that level is codified in acquisition regulations that limit the government's ability to budget for anticipated yet unspecified needs.

[9]. See Boeing's "World Air Cargo Forecast 2006–2007."

In: Civil Reserve Air Fleet
Editor: Soren M. Jonsson
ISBN: 978-1-62808-782-6

Chapter 3

MILITARY AIRLIFT: DOD SHOULD TAKE STEPS TO STRENGTHEN MANAGEMENT OF THE CIVIL RESERVE AIR FLEET PROGRAM*

United States Government Accountability Office

WHY GAO DID THIS STUDY

To move passengers and cargo, the Department of Defense (DOD) must supplement its military aircraft with cargo and passenger aircraft from commercial carriers participating in the Civil Reserve Air Fleet (CRAF) program. Carriers participating in CRAF commit their aircraft to DOD to support a range of military operations. In the Fiscal Year 2008 National Defense Authorization Act, Congress required DOD to sponsor an assessment of CRAF and required GAO to review that assessment. GAO briefed congressional staff on its observations. As discussed with the staff, GAO further analyzed some of the issues identified in its review. This report assesses (1) the extent to which DOD has assessed potential risks to the CRAF program, and (2) the extent to which DOD's management of CRAF supports program objectives.

For this engagement, GAO reviewed DOD-sponsored CRAF study reports and interviewed study leadership. GAO also interviewed over 20 of 35 CRAF

---

* This is an edited, reformatted and augmented version of United States Government Accountability Office, Publication No. GAO-09-625, dated September 2009.

participating carriers that responded to a request for a meeting, DOD officials, and industry officials.

What GAO Recommends

GAO is recommending that DOD (1) conduct risk assessments on two CRAF passenger and cargo issues and (2) develop policies to strengthen its management of the CRAF program. In comments on a draft of this report, DOD disagreed with the first recommendation and agreed with the second.

What GAO Found

DOD needs to establish the level of risk associated with declining charter passenger capabilities and DOD's increased need to move very large cargo. Although DOD depends on CRAF charter passenger aircraft to move more than 90 percent of its peacetime needs, there has been nearly a 55 percent decline in this CRAF capacity since 2003. In addition, since 2003, DOD's large cargo movement needs have increased with the acquisition of over 15,000 Mine Resistant Ambush Protected vehicles. Since there are no U.S. commercial cargo aircraft capable of moving cargo this size into Iraq and Afghanistan, DOD is using foreign-owned carriers to assist its military aircraft in such movements. However, there are scenarios where foreign-owned carriers may be unwilling or not allowed to fly. As a result, the lack of a commercial U.S. outsized cargo capability might restrict DOD's ability to meet its large cargo airlift needs in a timely manner. DOD has not quantified the risks these challenges pose to the CRAF program's ability to meet DOD's future transportation requirements because DOD has not completed risk assessments as described in the 2008 National Defense Strategy. Until risk assessments are conducted, DOD will not be sufficiently informed about potential risks in the CRAF charter passenger segment and in very large cargo airlift capability that could prevent DOD from managing its future airlift needs and the CRAF program effectively.

DOD's management of CRAF has not provided CRAF participants with a clear understanding, which could strengthen the program's ability to support its objectives, in some critical areas of the program. Although internal controls such as policies can help meet program objectives, CRAF business partners do not have a clear understanding of DOD's expectations concerning four CRAF objectives—an enhanced mobilization base, modernization, increased air carrier participation, and communication—because DOD has not developed policies in these four areas. First, DOD has not developed policies regarding the enforcement of its business rules, such as the 60/40 rule that states that participants should fly only 40 percent of their total business for DOD. DOD does not consistently enforce this rule and this may decrease the mobilization base since it is difficult for carriers to size their fleets to meet DOD demands. Second, DOD has not developed policies or economic incentives that promote CRAF modernization and this may hinder CRAF carriers from modernizing their aircraft. Third, DOD has not developed policies regarding oversight of the distribution of its peacetime airlift business, the primary incentive to carriers for participating in CRAF. DOD has no involvement in this distribution, and the perceptions of some carriers that this process is unfair could ultimately reduce carrier participation in CRAF. Fourth, DOD has not developed policy concerning communication with the carriers on CRAF studies or proposed changes to the CRAF program. DOD has not always communicated with carriers prior to implementing changes or completing studies. Until DOD develops policies that provide carriers with a clear understanding of CRAF, DOD cannot provide reasonable assurance that CRAF will meet its primary objective of providing critical airlift.

ABBREVIATIONS

AMC	Air Mobility Command
CRAF	Civil Reserve Air Fleet
DOD	Department of Defense
IDA	Institute for Defense Analyses
MRAP	Mine Resistant Ambush Protected

September 30, 2009

Congressional Committees

Since September 11, 2001, the U.S. Transportation Command's Air Mobility Command (AMC) has successfully transported more than 6.8 million passengers and nearly 3 million tons of cargo by air. To move these passengers and cargo, the Department of Defense (DOD) must supplement its military aircraft with aircraft from commercial carriers participating in the Civil Reserve Air Fleet (CRAF) program. Commercial air carriers participating in CRAF contractually commit their aircraft to DOD to be called upon, or activated, for use to support a range of military operations. At the most demanding end of this range, DOD plans for CRAF to move more than 90 percent of its passengers and almost 40 percent of its cargo requirements. As an incentive to encourage participation in CRAF, DOD contracts exclusively with CRAF participants to fly its daily peacetime passenger and cargo airlift business. Currently, this daily peacetime business includes airlift for operations in Iraq and Afghanistan, which is being handled without an activation of CRAF. CRAF has been activated twice—once during Operation Desert Shield/Storm and at the beginning of Operation Iraqi Freedom. The Office of the Under Secretary of Defense for Acquisition, Technology and Logistics (Transportation Policy) is responsible for establishing policies for the CRAF program. The U.S. Transportation Command is responsible for the daily management of the program and for making recommendations concerning the capability, capacity and other requirements for mobility assets needed to execute its mission.

The National Airlift Policy confirms the importance and the necessity of CRAF by establishing that national objectives cannot be met without using commercial air carriers.[1] DOD has little capability to meet its passenger airlift requirements in its military fleet; therefore, DOD relies almost entirely on CRAF commercial carriers to fulfill its requirements. Charter passenger carriers, operating on the customers' schedules, provide nearly 90 percent of all daily, peacetime passenger airlift for DOD, while the scheduled carriers, operating on set routes and timetables, provide the remainder. In contrast, DOD's military fleet has the capability to carry bulk, outsized, and oversized cargo;[2] however, DOD relies on commercial cargo carriers to move bulk cargo in both peacetime and times of crisis or war. Although CRAF cargo carriers can fly oversized cargo when needed, they do not have the capability to move outsized cargo. CRAF carriers are not capable of loading or moving strategic outsized equipment, such as Mine Resistant Ambush Protected (MRAP) vehicles, tanks, and helicopters.

Multiple DOD-sponsored studies have identified risks to the CRAF program, with some risks due to a changing business environment and challenges within the management of the CRAF program.[3] For example, as required by the Fiscal Year 2008 National Defense Authorization Act, DOD sponsored an independent assessment of the viability of the CRAF program. DOD submitted this assessment to Congress in October 2008, and as required by that act, we subsequently conducted a review of that assessment and briefed congressional staff on our observations about the assessment and some issues we identified about the CRAF program in general. As discussed with congressional staff, we further analyzed some of the issues identified in that review. This report assesses (1) the extent to which DOD has assessed potential risks to the CRAF program and (2) the extent to which DOD's management of CRAF supports program objectives.

To assess the extent to which DOD has assessed potential risks to the CRAF program and the extent to which DOD's management of CRAF supports program objectives, we reviewed several DOD-sponsored CRAF reports. We interviewed study leadership from the DOD-sponsored studies to obtain their perspectives and identify relevant issues relating to the CRAF program from their respective studies. We also conducted structured interviews with over 20 CRAF air carriers of the 35 participating in the program as of October 2008 that responded to our request for a meeting and interviewed airline industry officials and DOD officials at the Office of Secretary of Defense, U.S. Transportation Command, and Air Force Air Mobility Command to discuss the CRAF program. We analyzed this information in the context of DOD's National Airlift Policy, 2008 National Defense Strategy, and airlift requirements to determine how issues we identified might affect the CRAF program. We discussed the management of the CRAF program with officials at the Office of Secretary of Defense, U.S. Transportation Command, and Air Mobility Command. We also reviewed the 2008 National Defense Strategy, GAO's Standards for Internal Control in Federal Government,[4] DOD's Manager's Internal Controls, and U.S. Transportation Command's Manager's Internal Controls (guidance to implement DOD's internal controls program) to determine how programs should be managed to minimize the risk of program failure and to identify elements of effectively managed programs to determine if these elements are part of the CRAF program.

We conducted this performance audit from January 2009 to July 2009 in accordance with generally accepted government auditing standards. Those standards require that we plan and perform the audit to obtain sufficient, appropriate evidence to provide a reasonable basis for our findings and conclusions based on our audit objectives. We believe that the evidence obtained provides a reasonable basis for our findings and conclusions based on our audit objectives.

RESULTS IN BRIEF

DOD has not assessed the level of risk that declining charter passenger capabilities and DOD's increased need to move outsized cargo might have on future airlift requirements and the CRAF program. Although DOD depends on CRAF charter passenger aircraft to move more than 90 percent of its peacetime and contingency airlift needs, there has been nearly a 55 percent decline in this capability since 2003. In 2003, DOD relied on a charter air carrier capability comprised of more than 60 aircraft; however, although DOD passenger needs have increased significantly since 2003, the CRAF charter passenger air carrier capability has decreased, declining to as few as 19 aircraft in April 2008 before stabilizing at 29 aircraft in May 2008. The decline to 19 aircraft delayed the return home of Maine National Guard troops in Iraq by about a week and caused DOD to quickly find additional airlift. Because scheduled carriers have historically been unwilling and/or unable to fly the CRAF missions completed by charter carriers, this decline suggests that, at some point in the future, charter passenger carriers or aircraft participating in the CRAF program may not be able to meet DOD's airlift needs on a daily basis and in contingency operations. In addition, since 2003, DOD's need to move outsized cargo has increased with the acquisition of over 15,000 MRAP vehicles. As there are no U.S. commercial cargo aircraft capable of moving outsized cargo such as these high priority vehicles into Iraq and Afghanistan, DOD is using foreign-owned carriers to support such movements. According to DOD officials, using foreign-owned carriers relieves some of the stress on military aircraft—there are sufficient military aircraft to accomplish the missions—and is less expensive than military airlift. As the need to move outsized cargo has increased, use of foreign-owned carriers to move these assets has increased; however, in some scenarios foreign-owned carriers may be unwilling to fly, and in other scenarios, like a CRAF activation, foreign-owned carriers would only be used in exceptional circumstances. As a result, the lack of a U.S. outsized cargo capability might affect DOD's ability to meet its outsized cargo airlift needs in a timely manner. DOD has not established whether it will need to continue to use these foreign-owned aircraft or whether there is a need for a U.S. commercial capability to help relieve stress on the military aircraft, and has not completed risk assessments, as set out in the 2008 National Defense Strategy. The 2008 National Defense Strategy describes the need to assess, mitigate, and respond to risk in the execution of defense programs critical to national security. Until specific risk assessments are conducted and actions are taken to mitigate any risks that may exist in the

CRAF charter passenger segment and in outsized cargo capability that could prevent DOD from managing its future airlift needs and the CRAF program effectively, DOD and congressional decision makers will not be fully informed about unmitigated risks—specifically, gaps, shortfalls, or redundancies.

DOD's management of the CRAF program has not provided CRAF air carrier participants with a clear understanding of some critical areas of the program. Management internal controls help provide reasonable assurance that programs are focused on and can achieve their objectives by requiring clearly articulated policies. The National Airlift Policy states that federal policies governing the CRAF program should enhance the mobilization base,[5] promote aircraft modernization, increase air carrier participation, and provide a framework for dialogue and cooperation with commercial air carriers; however, we found that CRAF business partners do not have a clear understanding of DOD's expectations concerning these objectives because DOD lacks formal, written policies in four critical areas of the program, which could strengthen the program's ability to support its objectives.

- First, DOD's lack of policy regarding the enforcement of its business rules may negatively affect the CRAF mobilization base. DOD does not routinely enforce a basic CRAF business rule, outlined in the CRAF solicitation, allowing that no more than 40 percent of a CRAF carrier's revenues should come from the government. DOD has no formal policy describing this rule or when it will be enforced, thus leaving CRAF participants and those interested in becoming participants potentially unaware of DOD's intent. Also, CRAF carriers have stated that the inconsistent enforcement of this rule makes it difficult since leasing or buying aircraft to meet DOD needs requires long-term planning.
- Second, DOD has not developed CRAF policies that promote modernization, although modernization of the CRAF fleet is an objective of the National Airlift Policy and has been acknowledged as a goal by senior Air Force officials. Despite repeated congressional testimony from DOD officials stressing the importance of modernization, DOD does not provide CRAF participants any guidelines, objectives, or economic incentives that would encourage modernization. Some charter passenger air carriers, which help fly the majority of peacetime missions for DOD, fly aircraft in the Federal Aviation Administration's National Aging Aircraft Research Program.

In the absence of specific modernization objectives, CRAF business partners will remain unaware of DOD's expectations toward modernization, and might not take steps needed to modernize their aircraft.

- Third, DOD's lack of policy regarding the distribution of DOD's peacetime airlift business may negatively affect CRAF air carrier participation. According to DOD officials, the process and procedures for distributing DOD's peacetime airlift business among carriers have not been defined by and are not overseen by DOD. While DOD defines how participants earn entitlements to peacetime airlift business, DOD actually distributes the entitlements to three teams of CRAF participants. The teams then distribute the business based on arrangements made by each team's members. In our discussions with carriers, some carriers expressed concerns that the distribution of DOD's peacetime business was unfair, because carriers were not getting the distribution they expected based on their contributions to CRAF, and lacked transparency. Because the distribution of this business is intended to incentivize CRAF participation, this perception could ultimately reduce carrier participation in CRAF.
- Fourth, DOD has not developed policy that establishes a framework for dialogue and cooperation with commercial air carriers, which could involve CRAF participants in decision-making and facilitate sharing information with them. According to airline officials, DOD has not routinely involved CRAF carriers in proposed changes to the CRAF program in advance and it has not shared CRAF study results with the carriers until recently. As a result, CRAF participants told us that they have been unable to adjust their business plans to better address DOD needs and easily absorb some of the program changes that have been announced. DOD has told us that there are many ways in which they communicate with the carriers.

By developing policies that provide commercial air carriers with a clear understanding of critical aspects of the CRAF program, DOD can strengthen its management of the CRAF program and have reasonable assurance that the CRAF program will meet its primary objective of providing critical airlift to support DOD's potential future needs.

To provide additional assurance the CRAF program can continue to meet DOD airlift needs and to assist DOD in more effectively managing the program, we are making two recommendations. First, to help DOD identify and analyze risks that could assist DOD in meeting its future airlift requirements, we recommend that DOD conduct risk assessments that evaluate the declining U.S. charter passenger capability and the lack of a U.S. commercial outsized cargo capability and develop appropriate policies and procedures for mitigating any identified risks. Second, to facilitate effective management of the critical partnership between DOD and the U.S. commercial air carrier industry and to help ensure that CRAF can meet its program objectives, we recommend that DOD develop formal, written policy that establishes: enforcement guidelines for the basic CRAF business rules; objectives and measures of effectiveness required to achieve modernization of the CRAF fleets; oversight over how peacetime airlift business is distributed; and mechanisms by which DOD seeks input from CRAF participants in program issues and concerns and information sharing.

In commenting on a draft of this report, DOD did not agree with our recommendation to conduct risk assessments that evaluate the declining U.S. charter passenger capability and the lack of a U.S. commercial outsized capability. DOD commented that the 2008 CRAF viability study conducted by the Institute for Defense Analyses (IDA) evaluated the risk of the charter passenger industry, and based on that study, they are examining potential mitigation strategies for the declining passenger charter fleet. However, during our review, we found no evidence that U.S. Transportation Command has conducted an up-to-date risk assessment or is developing policies and procedures to mitigate any risks associated with declining charter passenger capability. As we reported, the 2008 IDA study data ended in 2007. Also, the study did not establish a level of acceptable risk for the CRAF program. Additionally, DOD commented that there was no need to develop any mitigation strategies for an outsized cargo capability because CRAF is not intended to provide an outsized cargo capability. Given DOD's recent addition of over 15,000 MRAPs, we continue to believe that risk assessments, as outlined in the National Security Strategy, of both the charter passenger and outsized cargo issue would give DOD specific information that would help them shape their future strategic transportation requirements. DOD concurred with our recommendation to develop formal, written policy that establishes: enforcement guidelines for the basic CRAF business rules; objectives and measures of effectiveness required to achieve modernization of the CRAF fleets; oversight over how peacetime airlift business is distributed; and mechanisms by which DOD seeks input from CRAF participants in program issues and concerns and information sharing. A more detailed discussion of

DOD's comments and our response to these comments follow the Recommendations for Executive Action section of this report.

BACKGROUND

The CRAF program was created in 1951 and its importance was reaffirmed by the National Airlift Policy in 1987. The National Airlift Policy states that the military will rely on the commercial air carrier industry to provide the airlift capability required beyond that available in the military airlift fleet. Additionally, the policy includes nine guidelines to meet airlift requirements in peacetime and wartime. These guidelines direct that policies be designed to

- increase participation in CRAF;
- enhance the mobilization base of the U.S. commercial air carrier industry;[6]
- provide a framework for dialogue and cooperation with commercial air carriers; and
- promote the development of technologically advanced transport aircraft and related equipment. According to DOD officials, these guidelines serve as the objectives of the CRAF program.

CRAF commitments are divided into three levels or stages—Stages I, II, and III—depending on the size of the operations or contingency in which DOD is involved. As defined in the CRAF contract between DOD and its commercial partners, Stage I activation supports expanded operations beyond DOD's routine daily operations and provides the equivalent of 30 passenger and 30 cargo aircraft; Stage II activation is used in the event of a major regional contingency and supporting mobilization and provides the equivalent of 87 passenger aircraft and 75 cargo aircraft; and Stage III activation supports two major regional contingencies and provides the equivalent of 136 passenger aircraft and 120 cargo aircraft. When CRAF is activated, carriers have a specified time frame in which to provide aircraft, with pilots and crews, to DOD. Once activated, air carriers continue to operate and maintain the aircraft with their resources; however, AMC controls the aircraft missions.

The majority of DOD passenger flights require carrier flexibility, as many DOD missions are not routine in their locations or timing. Charter passenger carriers fly the majority of DOD peacetime, contingency, and Stage I business because charter passenger carriers' businesses are designed with the flexibility to provide airlift based on the customer's (DOD's) schedule. Scheduled passenger carriers operate commercial flights on regular routes and can ill afford unplanned disruptions to their airline networks. However, because of their large fleet sizes, the scheduled carriers are a critical component of the CRAF fleet.

As an incentive to encourage participation in CRAF, DOD contracts exclusively with CRAF participants to fly its daily, peacetime passenger and cargo airlift business and any surge for contingencies. As articulated in an August 2008 DOD-sponsored CRAF study, carriers earn the entitlement to DOD's peacetime business through points awarded based on their aircraft commitments to each CRAF stage.[7] According to the study, the greater the commitment by the carrier, the greater the amount of peacetime business to which a CRAF participant is entitled. These points become the basis of a carrier's entitlement to compete for the procurement of peacetime passenger and cargo airlift business.

To maximize the value of these entitlements, CRAF participants have formed into three teaming arrangements, which are created and managed by the participants themselves. These teams comprise of a mix of passenger and cargo carriers that join together to pool their entitlements to DOD business; that is, the entitlement directly associated with a carrier's individual commitment is combined with the entitlements earned by other carriers on their team. DOD assigns peacetime business to the team based on the team's total entitlement and availability, not to the individual carrier. Once that business is assigned to the team, the team leader, or administrator, is responsible for accepting and distributing the business to the carriers at their discretion.

DOD's 2008 National Defense Strategy requires the military to assess, mitigate, and respond to risk that could potentially damage national security.[8] Identifying and managing risk is also an important goal of all successful internal control programs. Internal controls include the organization, policies, and procedures used by agencies to reasonably ensure that, among other things, critical programs like CRAF achieve their intended results effectively and efficiently. Internal control standards require that management should provide for an assessment of the risks the agency faces from both external and internal sources. These standards also require that there be control activities—that is, the policies, procedures, techniques, and mechanisms that enforce management's directives—in place to help insure that actions are taken to address risk.

DOD HAS NOT ASSESSED THE RISKS THAT CHANGES IN CHARTER PASSENGER CAPABILITIES AND DOD'S OUTSIZED CARGO NEEDS MIGHT HAVE ON THE CRAF PROGRAM

Charter Passenger Capability Has Declined

Although DOD depends heavily on CRAF charter passenger capability, this capability has declined substantially over the past 5 years and DOD has not established the risk that this decline may have for the CRAF program. DOD depends on the charter passenger industry to move more than 90 percent of its peacetime requirements, as well as all contingency surges. While the charter passenger capability has, historically, satisfied DOD's requirements, there has been a 60 percent decline in this capability since 2003 due mainly to a declining demand for charter airlines in the commercial sector. Figure I shows this decline in CRAF participants' charter passenger aircraft from a high of 66 aircraft in 2004 to 29 in 2008. Additionally, the figure shows that, even as commercial passenger carriers' revenues from DOD peacetime business increased to historic levels after 2001, and the amount of business available to charter passenger carriers was higher than it had ever been, the charter aircraft capacity continued to decline.

This decline in charter passenger capability led to a finding in the August 2008 DOD-sponsored CRAF viability study that this capability may become marginal for unexpected peacetime and contingency requirements. However, the study did not reflect that, in April 2008, CRAF's largest charter passenger carrier ceased operations due to bankruptcy. The sudden loss of 16 charter passenger airplanes from the CRAF program left DOD with only 3 charter passenger carriers and 19 total charter passenger aircraft until May 2008, when another passenger carrier dropped its scheduled services and committed 20 charter aircraft to CRAF. However, according to industry officials and confirmed by DOD, the sudden reduction in charter aircraft after the April 2008 bankruptcy led to a situation in which the return home of some redeploying Maine National Guard troops in Iraq was delayed by about a week. Because of a limited charter aircraft capability, the Commander, U.S. Transportation Command, personally called CRAF scheduled carriers and asked them to free up aircraft to transport these troops back to the United States.[9] The bankruptcy of DOD's largest charter passenger carrier without notice demonstrates the volatility of the charter passenger industry and raises questions about the industry's ability to continue to meet DOD requirements without a CRAF activation involving the larger, scheduled carriers to satisfy the requirements the charter passenger industry was filling.

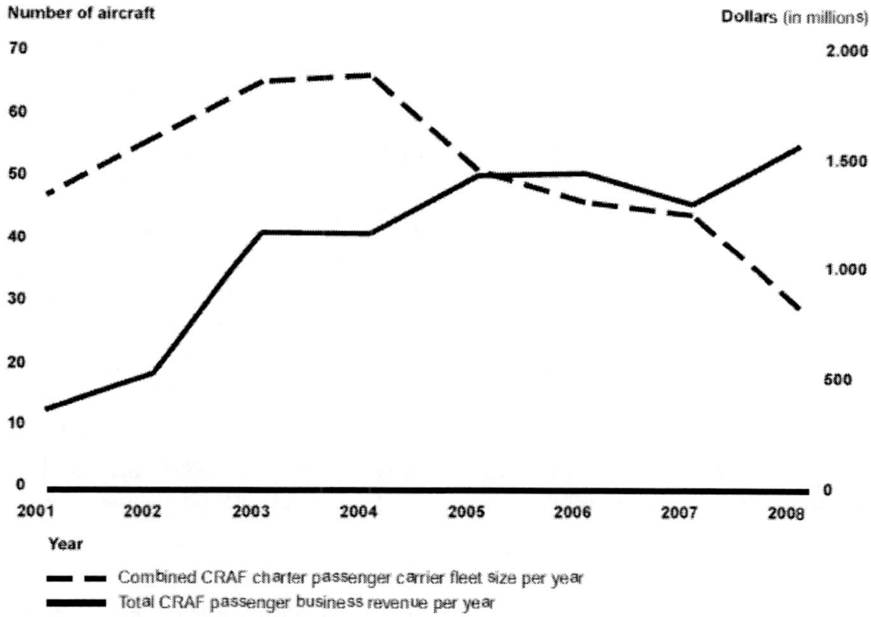

Source: GAO analysis of CRAF data.

Figure 1. Size of CRAF Charter Passenger Fleet and DOD Peacetime Business (Fiscal Years 2001 through 2008).

There is little or no excess capacity among scheduled carriers. The five scheduled passenger carriers we spoke with told us that, due to market conditions and shrinking fleets that have been tailored to meet their commercial demands more efficiently, scheduled carriers are reluctant to commit aircraft to peacetime operations, contingencies, and CRAF Stage I beyond a small, required contribution.[10] According to airline and industry officials, pulling a single aircraft out of a scheduled passenger carrier's daily planned service can cause major disruptions to its routes; therefore, to support any stage of CRAF activation, scheduled air carriers depend on a decrease in their commercial demands, similar to the reductions seen after September 11, 2001, that would make aircraft available. If the charter passenger industry business continues to decline, DOD will likely be forced to turn to scheduled air carriers to fulfill daily and Stage I requirements currently met by the charter carriers. However, given the scheduled carriers' smaller fleets, DOD has not quantified the number of charter passenger aircraft it may need on a daily basis and in contingencies and Stage I, or the risk of having a smaller charter passenger capability to handle these requirements. DOD officials have told us that they have no concerns that sufficient CRAF participants will respond to a call for airlift, whether during peacetime or in an activation.

DOD's Need to Move Outsized Cargo Has Increased

Since 2005, DOD's need to move outsized cargo to support peacetime and contingency operations has increased with the acquisition of more than 15,000 MRAP vehicles. Because there are no U.S. commercial cargo aircraft capable of moving outsized cargo such as MRAP vehicles into Iraq and Afghanistan, DOD is using foreign-owned carriers to support such movements to supplement its military airlift capability. As of April 2009, DOD had moved a total of 3,890 MRAPs by air, of which almost 80 percent were moved using foreign-owned carriers flying large Antonov-124 aircraft. We found and DOD confirmed that in the 2005 Mobility Capabilities Study, DOD planned for U.S. commercial cargo carriers participating in the CRAF program to move only bulk cargo, and did not identify a need for these carriers to move outsized cargo; however, without some supplemental capability—such as the use of foreign-owned carriers—the need for DOD to move outsized cargo into areas of crisis, and have that cargo arrive in a timely manner, could limit DOD's ability to meet future airlift requirements. According to DOD analysts involved in the ongoing Mobility Capabilities and Requirements Study—2016, DOD will again plan for CRAF cargo participants to carry only bulk cargo.

As DOD moves additional troops and equipment into land-locked Afghanistan and in similar scenarios in the future, the need to airlift MRAPs and other large equipment, like helicopters, may continue to need the use of commercial carriers to assist military airlift. However, it is not clear whether foreign-owned carriers would be able or willing to fly in certain scenarios. For example, DOD officials acknowledged that foreign-owned carriers, for security reasons, would not likely be used during a CRAF activation.[11] Moreover, we believe the use of foreign-owned companies in support of U.S. military operations could be problematic if or when foreign-owned carriers find supporting a U.S. contingency to be inconsistent with their national interests. For example, in 2008, when the U.S. Transportation Command was using Russian-based carriers to fly outsized cargo to Iraq, Afghanistan, and other locations, U.S. military aircraft ferried Georgian troops from Iraq back to Georgia in anticipation of a potential confrontation with Russian troops. We believe that risk may be increased in such scenarios in the future. Without further analysis of DOD's options for meeting its outsized cargo needs, including the potential role of commercial carriers, the inability of DOD to meet its needs to move outsized cargo into areas of crisis and have that cargo arrive in a timely manner could increase risk for DOD operations. However, DOD officials told us that DOD is using the foreign-owned aircraft only to ease the high stress on military aircraft and because such use is less expensive than military aircraft, not because there is an insufficient number of military aircraft available to fly this outsized cargo.

DOD Is Not Fully Aware of How Changes in Its Charter Passenger Airlift Capabilities and DOD's Outsized Cargo Needs Affected CRAF Because It Has Not Conducted Risk Assessments

DOD is not fully aware of the extent to which these changes may have affected the CRAF program's ability to meet DOD's future transportation requirements because DOD has not conducted risk assessments as described in the 2008 National Defense Strategy. In this strategy, DOD defines risk to the national defense in terms of the potential for damage to national security combined with the probability of occurrence and a measurement of the consequences should the underlying risk remain unaddressed. This strategy also states that DOD must account for future challenges and their associated risks[12] to meet the objective of winning our nation's wars and describes the need to assess, mitigate, and respond to risk in the execution of defense programs critical to national security. In the case of the CRAF program, risk assessments can be used to determine if there are any gaps, shortfalls, or redundancies in the charter passenger or outsized cargo segments that could prevent DOD from meeting future airlift requirements.

The most recent DOD sponsored CRAF study, issued in August 2008, predicted that passenger charter capability may become marginal, but the capabilities reviewed in the study did not include the further declines in this capability that occurred in 2008. The study also did not quantify the risk associated with the passenger charter capability decline that has already occurred. In accordance with both GAO and DOD management internal controls,[13] a risk assessment could inform program managers by establishing the maximum and minimum acceptable risk for the CRAF program. For example, it could identify the numbers of charter passenger aircraft necessary to meet DOD requirements. Without a risk assessment, DOD will continue to be uncertain what the levels of required CRAF charter participation is necessary to fulfill requirements, and DOD and industry decisions makers will not be able to begin to take steps to address the risks.

Furthermore, according to DOD officials, DOD has not conducted a risk assessment that examines outsized cargo movement, including the use of commercial air carriers to supplement its military fleet, and identifies any consequences of relying on foreign owned carriers to meet peacetime and contingency needs. As previously stated, DOD is using foreign-owned carriers to move MRAPs and other outsized equipment to Afghanistan and Iraq. However, the 2005 Mobility Capabilities Study[14] predates the acquisition of more than 15,000 outsized MRAPs. Additionally, the August 2008 CRAF study did not assess any CRAF outsized cargo movement. A risk assessment could determine whether a gap, shortfall, or redundancy exists in relation to the U.S. commercial and military outsized cargo capability. In addition, a risk assessment could provide information to decision makers regarding the possibility of potential damage to national security from the reliance on foreign-owned carriers and the probability of such damage in future contingencies. Without such a risk assessment, DOD may not know the most effective method for transporting outsized cargo, and if any methods present potential risk to national security.

DOD Has Not Issued Policies That Would Strengthen Management of the CRAF Program

DOD's management of the CRAF program has not provided CRAF air carrier participants with a clear understanding of some critical areas of the program, which could strengthen the program's effectiveness and the ability to support its objectives. Although management internal controls such as clearly articulated policies can help meet program objectives, DOD has not developed policies related to four of the CRAF program objectives as outlined in the National Airlift Policy. These four objectives include: enhancing the mobilization base,[15] promoting aircraft modernization, increasing air carrier participation, and providing a framework for dialogue and cooperation with commercial air carriers. As outlined by both GAO and DOD, management internal controls help provide reasonable assurance that, through effective management, programs can achieve their objectives. According to these management internal controls, one way to help assure that a program's objectives are met is to establish clearly articulated policies. Policies, a form of management control, are, according to U.S. Transportation Command, intended to provide guidance and procedures to carry out operations or achieve objectives. However, we found that CRAF business partners do not have a clear understanding of important aspects of the CRAF program because DOD lacks policies in critical areas of the CRAF program that could help DOD meet its program's objectives. U.S. Transportation Command officials have stated that the CRAF contract with the carriers serves as policy. However, the contract does not contain some elemental items of policy including objectives, goals, and measures of effectiveness as outlined in GAO and DOD management internal controls.

DOD Has Not Developed Policies Related to Four CRAF Program Objectives

DOD Has Not Developed Policy Concerning 60/40 Rule Enforcement

DOD has not developed policy regarding the enforcement of its business rules, such as the 60/40 rule, that would help strengthen the CRAF mobilization base. More than 40 years ago, DOD established measures to ensure that CRAF air carriers had both commercial and DOD revenue streams. These measures evolved into what is now known as the 60/40 rule, a rule defined in the CRAF solicitation allowing that no CRAF carrier should collect more than 40 percent of its revenues from DOD business. Carriers that earn more than 40 percent of their revenue from DOD may be penalized by reductions in their entitlement to DOD business. The original goals of the rule were to ensure that CRAF carriers maintained a strong business base, efficient operations, and modern fleets, all of which would prevent carriers from going out of business when DOD demands were low. The rule would also provide DOD with a surge capability to draw on if demand grew suddenly. Although DOD created the 60/40 rule with these intended goals, several CRAF carriers told us that they are unaware of the intent of the rule today because they are not sure if they have to follow the rule, or if it is even being enforced.

Some CRAF carriers have broken the 60/40 rule by depending in large part on DOD for their revenue. However, because there is no written DOD policy describing the rule and its enforcement, no carrier could tell us when, or under what conditions, the rule is actually enforced. According to airline officials, this lack of guidance affects carriers' business plans because they are not sure whether to account for 60/40 rule compliance when determining the size their fleets. Unclear enforcement parameters also make it difficult to plan lease or purchase of planes or how many to acquire. Three DOD-sponsored CRAF studies completed in the last 3 years have all given differing recommendations regarding the 60/40 rule, adding to the ambiguity as to whether or not the rule is or will be in effect. Additionally, it is unclear whether or not CRAF objectives of participation and meeting DOD surge demands are being met. Without policy that clearly states the guidelines and objectives of the 60/40 rule, CRAF carriers may not be able to properly size their fleets to meet DOD demands, and have the capacity for DOD to draw on to meet demands, which may decrease the mobilization base of the CRAF program.

DOD Has Not Developed Policy Concerning CRAF Fleet Modernization

DOD has not developed policies that promote CRAF fleet modernization, although DOD officials have recognized the need for a more modern CRAF fleet. The National Airlift Policy directs that policies be created to promote the development of technologically advanced transport aircraft in order to ensure a commercial airlift capability. In addition, a December 2007 DOD-sponsored CRAF study acknowledged the importance of modernization and recommended that DOD develop policies to encourage CRAF carriers to modernize their existing fleets. Moreover, DOD officials have recognized the necessity of a modernized commercial air fleet by repeatedly testifying before Congress about its importance for the continued viability of the program.

However, DOD has not provided CRAF participants with policies that include guidelines, objectives, or economic incentives that would encourage modernization. Because the charter passenger industry plays such a large role in moving DOD passengers, we believe it is in DOD's interest to ensure the commercial airlines have guidelines and incentives, such as a rate structure that would pay more for carriers to fly newer airplanes, to assist in modernizing their fleets. Two of DOD's largest remaining charter passenger carriers are flying large numbers of aircraft listed on the Federal Aviation Administration's Aging Aircraft List. As the December 2007 DOD-sponsored CRAF study warned, these planes will soon be retired as the costs of inspections, maintenance, and life-extension work becomes prohibitive. Since these aircraft are being used to fly DOD business almost exclusively, charter passenger carriers told us that they look to DOD to provide guidance and incentives to modernize. DOD officials told us that they cannot influence modernization or force the carriers to modernize. However, without DOD policy that provides specific modernization guidelines, CRAF carriers may not see a reason or have a business case to take steps needed to modernize their aircraft.

DOD Has Not Developed Policy Concerning Oversight of Distribution of Peacetime Business

DOD has not developed policies regarding the oversight of distribution of its peacetime airlift business, which may negatively affect CRAF air carrier participation and may affect DOD's ability to manage the CRAF program effectively. DOD's incentive system of contracting with CRAF participants to fly its daily peacetime business is intended to meet the program objective of increasing air carrier participation in CRAF by providing each CRAF participant with a reasonable share of peacetime business. DOD policy that includes guidance, instructions, regulations, procedures, or rules that clarify the CRAF incentive system and some oversight of the distribution of peacetime business would give CRAF carriers a clearer understanding of this important process.

According to DOD officials, the process and procedures for distributing DOD's peacetime airlift business have not been described in policy and are not overseen by DOD. As discussed earlier, DOD awards individual carriers points based on the number and type of aircraft they commit to CRAF. These points become the basis of a carrier's entitlement to compete for the procurement of peacetime passenger and cargo airlift business. To maximize these points, the carriers have formed themselves into three teams that have their own agreements on how the business will be distributed among the team members. U.S. Transportation Command officials confirmed that they distribute peacetime business to the teams and have no further involvement in how the teams distribute peacetime business among the members. The officials also said that they consider the existing system to be adequate in meeting program objectives. In the absence of DOD oversight and control, some of the CRAF carriers have expressed concerns that peacetime business distribution is not transparent and can be inequitable. Some CRAF participants have told us that teams distribute DOD peacetime contracts disproportionate to an individual air carrier's CRAF commitment. These carriers also told us that the result is that some CRAF participants receive less DOD business than their entitlement reflects. Some CRAF carriers told us that the execution of the incentive system discourages participation and, in some instances, could cause carriers to go out of business. We understand that U.S. Transportation

Command and Air Mobility Command have no involvement with the formation of the teams or the agreements teams have reached with their members. However, without DOD policies and oversight over the final distribution of the peacetime business that flows from the incentive system established by DOD, DOD cannot be sure that this system is accomplishing its goal of enhancing carrier participation in CRAF.

DOD Has Not Developed Policy Concerning DOD and CRAF Carrier Communication

DOD has not developed policy that establishes a framework for dialogue and cooperation with commercial air carriers that would invite CRAF participants to comment on pending program decisions and facilitate sharing information with them. Although facilitating an effective partnership between DOD and commercial carriers is a stated objective of the CRAF program, airline officials stated that DOD has not involved CRAF participants in some important program decisions that have had significant impact on the participants' business plans. For example, DOD announced a policy change that decreased the amount of money carriers were reimbursed for fuel, which is allowed under the CRAF contract. Carriers told us that they factor in fuel reimbursements in their yearly business plans, and are not prepared to adjust to a significant pricing change during the middle of a year, especially when they had no knowledge of the change ahead of its implementation and thus could not plan in advance for the effects. In addition, until recently, DOD had not shared information from DOD-sponsored studies on the CRAF program with CRAF carriers. For example, of the four DOD-sponsored CRAF studies completed in the past 6 years, most carriers we talked to told us they had neither seen nor heard of the studies until late 2008. Of the CRAF carriers we interviewed, only one carrier reported receiving a copy of any DOD-sponsored CRAF study.

DOD officials have said that mechanisms are in place for effective CRAF communication between DOD and CRAF carriers, such as using trade associations to perform what DOD officials describe as an "industrial reality check" and holding industry days and conferences. However, according to some carriers, communication through trade associations is not sufficient because some carriers are not allowed a voice in meetings, and some carriers are not members of the associations at all. Several carriers stated that DOD has little communication with them beyond using trade associations and annual meetings. With a clearly described policy that establishes a framework for an effective partnership fostering communication, DOD could strengthen its management of the CRAF program and enhance its relationship with the carriers, thus ensuring continued participation in CRAF.

CONCLUSION

Given the importance of CRAF in moving passengers and cargo for DOD to support peacetime and contingency operations and major operations requiring CRAF activation, it is critical for the CRAF program to be able to meet DOD's future needs. By policy, statute, and contract, DOD depends on CRAF business partners that increasingly find themselves in a challenging business environment. If the charter passenger industry continues to decline, DOD could increasingly turn to scheduled air carriers to fulfill the daily and Stage I requirements that are currently being met by the charter carriers; however, the scheduled carriers may not be willing or able to fly these missions and meet DOD's airlift needs. Additionally, the potential absence of sufficient outsized cargo capability could potentially jeopardize national security by preventing DOD from accomplishing its missions due to an inability to move outsized cargo into areas of crisis within the time frames the commanders need it to arrive.

Until risk assessments are conducted and actions are taken to mitigate any risks that are identified, DOD and industry decision makers will not be fully informed about risks in the CRAF charter passenger segment and in outsized cargo capability that could prevent CRAF from meeting DOD's airlift requirements. Moreover, the lack of appropriate policies that address critical areas of the CRAF program hinders DOD's ability to meet the objectives of the program.

Until DOD develops policies that provide commercial air carriers with a clear understanding of critical aspects of the CRAF program, such as enforcement of business rules (such as the 60/40 rule), specific modernization guidelines, distribution of peacetime business, and a framework for communication, thus strengthening its management of the program, DOD cannot provide reasonable assurance that the CRAF program will meet its primary objective of providing critical airlift to support DOD operations.

Recommendations for Executive Action

To assist DOD with management of the CRAF program, we are making the following two recommendations for executive action.

First, to help DOD identify and analyze risks associated with achieving program objectives, we recommend that the Secretary of Defense direct the Commander, U.S. Transportation Command to

- Conduct risk assessments as outlined in DOD's National Defense Strategy that
  - evaluate the declining U.S. charter passenger capability by establishing the maximum and minimum acceptable risk for the CRAF program expressed in terms of numbers of charter passenger aircraft necessary to meet DOD requirements; and
  - evaluate the lack of an outsized cargo capability to supplement military capability and the extent to which the reliance on foreign owned carriers is appropriate;
- Develop appropriate policies and procedures for mitigating any identified risks.

Second, to strengthen the effectiveness of the critical partnership between DOD and the U.S. commercial air carrier industry and the management of the CRAF program to achieve its objectives, we recommend that the Secretary of Defense direct the Under Secretary of Defense for Acquisition, Technology and Logistics (Transportation Policy) to develop policy that

- establishes enforcement guidelines for the basic CRAF business rules, to include intent, objectives, and measures of effectiveness mechanisms;
- establishes incentives, objectives and measures of effectiveness required to ensure modernization of the CRAF fleets;
- establishes and describes oversight mechanisms by which DOD will monitor how peacetime airlift business is distributed to ensure that its CRAF incentive program is working as intended; and
- establishes and describes the mechanisms by which DOD includes CRAF participants to provide comments on pending program decisions and in information sharing, and that includes objectives and measures of effectiveness of these activities.

AGENCY COMMENTS AND OUR EVALUATION

In written comments on a draft of this report, DOD did not agree with our first recommendation to conduct risk assessments regarding the declining charter passenger capability and the lack of an outsized cargo capability as part of the Civil Reserve Air Fleet, but partially agreed with that part of the recommendation to develop policies and procedures to mitigate any identified risks. DOD agreed with our second recommendation to develop policy for aspects of the CRAF program.

While DOD disagreed with our recommendation to conduct risk assessments, DOD agreed with the value of conducting a risk assessment on the declining U.S. charter passenger charter capability, stating that this has already been evaluated by the CRAF viability study conducted by IDA DOD also stated that, based on the recommendations of the IDA study, DOD is already examining the declining passenger charter fleet and potential mitigation strategies. However, as we stated in our report, IDA's report included data that stopped at 2007 and did not include data regarding the 2008 business termination of a carrier that provided nearly 50 percent of the charter passenger capability available to DOD. Also, while the IDA report stated that the charter passenger industry may become marginal, data analysis that supported this statement did not establish the maximum and minimum acceptable risk for the CRAF program. Therefore, we continue to believe that our recommendation to establish acceptable risk levels is still viable and important.

DOD also disagreed with the second part of our recommendation concerning the need to conduct a risk assessment on the lack of a CRAF outsized cargo capability, stating that the CRAF program is not intended to provide outsized cargo capability. In their comments, DOD stated that its use of foreign carriers to transport outsized cargo is a strategy to reduce costs, save military flying hours and flight crews for higher priority missions, reduce military footprint, or provide flexible contract length/timing. DOD also stated that it is not an indication of a shortfall in the DOD outsize cargo capability or the CRAF program. However, as we reported, DOD used foreign-owned carriers flying AN-124 aircraft to move high priority outsized cargo (MRAPs) into Iraq instead of the organic fleet of C-5s and C-17s. We did not state that there was a shortfall in either the CRAF program or DOD outsized capability. Rather, we point out that, if DOD is to know whether there is a shortfall, gap, or redundancy in that capability, particularly given the addition of over 15,000 MRAPs, they would need to do a risk assessment. We continue to believe that a risk assessment of this issue would give DOD specific information that would help it shape future strategic transportation requirements.

DOD partially agreed with the third part of our recommendation pertaining to the need to develop appropriate policies and procedures for mitigating any identified risks regarding the decline of charter passenger capability and lack of outsized cargo capability. DOD stated that U.S. Transportation Command is examining potential mitigation strategies for the declining U.S. passenger charter segment. However, during our review, we found no evidence that U.S. Transportation Command was developing policies and procedures to mitigate any risks associated with declining charter passenger capability and outsized cargo capability. DOD disagreed with the need to develop any mitigation strategies for an outsized cargo capability since CRAF is not intended to carry outsized cargo. As stated above, DOD's use of foreign-owned carriers to move outsized MRAPs would lead us to believe that there might be a future need for policies and procedures to mitigate any shortfall or gap.

DOD agreed that there is a need for comprehensive policy governing all of the CRAF program elements identified in our draft report. However, DOD did not identify what, if any, specific actions it would take in response to our recommendation. We encourage DOD to establish enforcement guidelines for CRAF business rules; objectives and measures of effectiveness for modernization; oversight mechanisms describing how peacetime business should be distributed; and mechanisms for information sharing.

William M. Solis
Director,
Defense Capabilities and Management

APPENDIX I: RECENT DEPARTMENT OF DEFENSE STUDIES

Sustaining the Civil Reserve Air Fleet (CRAF) Program, Institute for Defense Analyses, May 1, 2003.

Economic Review of the Civil Reserve Air Fleet (CRAF) Program, Institute for Defense Analyses, December 15, 2007.

Civil Reserve Air Fleet (CRAF) Study Report, Council for Logistics Research, July 13, 2008.

Civil Reserve Air Fleet: Economics and Strategy, Institute for Defense Analyses, August 22, 2008.

End Notes

[1] National Security Decision Directive 280, June 24, 1987.

[2] There are three designations of cargo types: bulk, oversized, and outsized. Bulk cargo is cargo that will fit on a 463L cargo system pallet or standard commercial airline pallet/cargo container. Oversized cargo is air cargo that exceeds the dimensions of a standard (463L) pallet, but is air transportable on military and most civilian cargo carriers. Outsized cargo exceeds the dimensions of oversized cargo and requires the use of a military C-5 or C-17 aircraft or surface transportation.

[3] See appendix I for a list of the studies.

[4] GAO, *Standards for Internal Control in the Federal Government*, GAO/AIMD-00-21.3.1 (Washington, D.C.: November 1999).

[5] DOD defines the mobilization base as the total of all resources available, or that can be made available, to meet foreseeable wartime needs. Such resources include the manpower and materiel resources and services required for the support of essential military, civilian, and survival activities, as well as the elements affecting their state of readiness, such as planning with industry and modernization of equipment.

[6] DOD defines the mobilization base as the total of all resources available, or that can be made available, to meet foreseeable wartime needs. Such resources include the manpower and materiel resources and services required for the support of essential military, civilian, and survival activities, as well as the elements affecting their state of readiness, such as planning with industry and modernization of equipment.

[7] Upon commitment and acceptance to the CRAF program, the CRAF carrier's aircraft are assigned points based on range, payload, and productive utilization rate, essentially determining the value of each aircraft to the program.

[8] The National Defense Strategy defined risk in terms of the potential for damage to national security combined with the probability of the occurrence and the measurement of the consequences should the underlying risk remain unaddressed.

[9] It should be noted that the teaming arrangements may provide DOD with an additional means to enforce the contract. Under the terms and conditions of some of these contracts the carriers are jointly and severely liable and would be contractually bound to provide the aircraft and services required under the terms and conditions of the contract.

[10] All carriers must commit at least one aircraft to Stage I to be eligible to commit to the later stages.

[11] While foreign-owned carriers cannot fly CRAF missions, foreign-owned carriers can fly DOD (CRAF-like) chartered missions during peacetime operations in compliance with the Fly CRAF Act. These DOD chartered missions are contracted to CRAF participants, which subcontract these missions to foreign-owned carriers.

[12] Future challenges risks are those associated with the department's capacity to execute future missions successfully against an array of prospective future challengers.

[13] GAO, *Internal Control Management and Evaluation Tool,* GAO-01-1008G (Washington, D.C.: August 2001); GAO, *Standards for Internal Control in the Federal Government,* GAO/AIMD-00-21.3.1 (Washington, D.C.: November 1999); and Department of Defense Instruction 5010.40, Managers' Internal Control Program Procedures (January 4, 2006).

[14] The Mobility Capabilities Study which began in May 2004, reassessed DOD's mobility capabilities against the backdrop of a revised National Military Strategy. The study results were intended to support decisions on future strategic airlift needed to meet varying military

requirements of combat operations and smaller military efforts. This study is the fourth in a series of major mobility studies that DOD has conducted since the end of the Cold War.

[15] DOD defines the mobilization base as the total of all resources available, or that can be made available, to meet foreseeable wartime needs. Such resources include the manpower and materiel resources and services required for the support of essential military, civilian, and survival activities, as well as the elements affecting their state of readiness, such as planning with industry and modernization of equipment.

In: Civil Reserve Air Fleet
Editor: Soren M. Jonsson
ISBN: 978-1-62808-782-6

Chapter 4

MILITARY AIRLIFT: DOD NEEDS TO TAKE STEPS TO MANAGE WORKLOAD DISTRIBUTED TO THE CIVIL RESERVE AIR FLEET*

United States Government Accountability Office

WHY GAO DID THIS STUDY

To move passengers and cargo, DOD supplements its military aircraft with cargo and passenger aircraft from volunteer commercial carriers participating in the CRAF program. Participating carriers commit their aircraft to support a range of military operations in exchange for peacetime business. A House Armed Services Committee mandated GAO to report on matters related to the CRAF program. GAO assessed whether DOD (1) met its military airlift training requirements while also using CRAF participants to the maximum extent practicable, (2) provided justification for restricting commercial carriers from transporting partial plane loads of cargo over certain routes, and (3) has established future requirements for CRAF and how the planned size of CRAF compares to those requirements. GAO reviewed guidance and policies pertaining to the program, flying hour data, and DOD-

* This is an edited, reformatted and augmented version of United States Government Accountability Office, Publication No. GAO-13-564, dated June 2013.

sponsored CRAF study reports. GAO also interviewed DOD and industry officials.

What GAO Recommends

GAO recommends that the Secretary of Defense direct the Secretary of the Air Force and the Commander, U.S. Transportation Command—in conjunction with the Commander, Air Mobility Command—to use its existing processes for monitoring training to determine when it can shift its distribution of peacetime airlift workload from military to commercial sources. In comments on a draft of this report, DOD concurred with GAO's recommendation and stated that it believes implementing the recommendation will further improve the Civil Reserve Air Fleet program.

What GAO Found

DOD exceeded the flying hours needed to meet military training requirements for fiscal years 2002 through 2010 because of increased operational requirements associated with Afghanistan and Iraq; however it does not know whether it used Civil Reserve Air Fleet (CRAF) participants to the maximum extent practicable. DOD guidance requires it to meet training requirements and to use commercial transportation to the "maximum extent practicable." During fiscal years 2002 through 2010, DOD flew its fleet more than needed to train its crews, although its flying has more closely matched its training needs in recent years. DOD has also used CRAF participants extensively to supplement military airlift. Although DOD has taken steps to make more airlift business available to CRAF participants, officials said that overseas operations have provided enough missions to support both training and CRAF business obligations. However, with the drawdown in Afghanistan, DOD officials expect the need for airlift to decline by at least 66 percent—to pre-September 2001 levels—reducing both training hours available for DOD and business opportunities for CRAF. DOD does not use its process for monitoring flying hours to determine when it will exceed required training hours and allocate eligible airlift missions to CRAF participants. Therefore, it cannot determine whether it is using CRAF to the maximum extent practicable. As a result, DOD may be using its military fleet more than necessary—which officials say is less economical—while risking reduced CRAF participation.

DOD provided several reasons for restricting commercial carriers from transporting partial plane loads of cargo over channel routes, including the need to promote efficiency, meet its military airlift training requirements, and fulfill peacetime business obligations to CRAF participants. Channel route missions are regularly scheduled airlift missions used to transport cargo and provide aircrew training time. These missions also help DOD provide business to CRAF participants. According to U.S. Transportation Command (TRANSCOM) officials, DOD generally requires aircraft conducting channel route missions to be completely full of cargo before takeoff. The policy restricting carriers from flying partial loads over channel routes allows DOD to consolidate cargo previously flown by commercial carriers in less than full plane loads and redirect that cargo into the channel route system, where it will be transported by either commercial or military aircraft as part of a full plane load mission. According to DOD, consolidating cargo into full loads flown over the channel route system has increased both the efficiency of these missions and the availability of missions that DOD uses to train its crews and fulfill its business obligations to CRAF.

It is unclear whether the planned size of CRAF will be adequate to meet future airlift requirements. DOD last established its future requirements based on the wartime scenarios in the Mobility Capability Requirements Study 2016, issued in 2010. However, due to changing military strategy and priorities, the 2010 study does not reflect current mission needs. The National Defense Authorization Act for Fiscal Year 2013 requires DOD to conduct a new mobility capabilities and requirements study. DOD has not begun this study or finalized its ongoing reviews of the CRAF program's ability to support future requirements. Once they are finalized, these studies should allow DOD to better understand future requirements for CRAF and whether the CRAF program will meet future airlift requirements.

ABBREVIATIONS

AMC	Air Mobility Command
CRAF	Civil Reserve Air Fleet
DOD	Department of Defense
MCRS	Mobility Capability Requirements Study
MRCS	Mobility Requirements and Capability Study

NDAA	National Defense Authorization Act
REMIS	Reliability and Maintainability Information System
TACC	Tanker Airlift Control Center
TRANSCOM	U.S. Transportation Command

441 G St. N.W. Washington, DC 20548
June 20, 2013

Congressional Committees

DOD augments its military cargo and passenger airlift capability with aircraft from commercial carriers participating in the Civil Reserve Air Fleet (CRAF) program. Participating carriers commit their aircraft to be called upon, or activated,[1] for use to support a range of military operations. The U.S. government provides commercial carriers with economic incentives to encourage their participation in CRAF, by giving them the opportunity to transport DOD personnel and cargo in peacetime and by reserving a number of additional airlift contracts for CRAF carriers, such as the U.S. General Services Administration's City Pairs program and the U.S. Transportation Command-managed World Wide Express program. Since 2001, CRAF participants have received nearly $30 billion in revenue for providing peacetime airlift services to DOD.[2] In fiscal year 2012, CRAF participants received approximately $2 billion in payments from DOD for their peacetime participation. However, DOD projects that by 2015, its peacetime business will decline by an estimated 66 percent as it returns to pre-September 11, 2001 levels, raising concerns within DOD and among CRAF participants about the effect this loss of business will have on the carriers and their ability to continue participating in the program. The Office of the Deputy Assistant Secretary of Defense for Transportation Policy serves as the principal advisor for establishing policies and providing guidance to DOD components for efficient and effective use of government and commercial transportation resources. DOD's U.S. Transportation Command (TRANSCOM), headquartered at Scott Air Force Base in Illinois, manages many aspects of the CRAF program and makes recommendations concerning the capability, capacity, and other requirements for mobility assets needed to execute TRANSCOM's mission.

According to TRANSCOM officials, CRAF benefits the nation by providing up to half of the nation's long-range airlift capability, allowing the government to avoid having to buy additional aircraft, pay personnel costs, or maintain the aircraft during peacetime. The National Airlift Policy emphasizes the importance of CRAF to the nation by noting that military and commercial resources are equally important and interdependent in the fulfillment of the national defense airlift objective.[3] Consistent with that policy, DOD Instruction 4500.57 requires that DOD operate its fleet to meet its training requirements while also using commercial sources of transportation to the "maximum extent practicable."[4] According to Air Mobility Command (AMC) officials, airlift missions that are not conducted to satisfy training requirements should be performed by CRAF participants, except when there is some other feature of the mission that requires military airlift. Further, since DOD's military fleet has limited capability to meet its passenger airlift requirements, it mostly relies on CRAF commercial carriers to fulfill these requirements. DOD's military fleet has the capability to carry bulk, outsized, and oversized cargo; however, DOD relies on commercial cargo carriers to move a portion of its bulk cargo both in peacetime and in times of crisis or war.

Over the last several years, we and others have reported on the challenges faced by the CRAF program. In September 2009, we recommended that DOD assess the level of risk associated with the commercial carriers' declining charter passenger capabilities and DOD's increased need to move very large cargo. DOD did not agree with our recommendation because, it stated, an assessment of the declining charter passenger capabilities had already been done in 2008 by the Institute for Defense Analysis; however, as we stated in our report, the assessment conducted by the Institute for Defense Analysis on DOD's behalf did not include the most current and relevant data affecting the program; for that reason we believe our recommendation is still viable and important. Additionally, we recommended that, given the importance of the CRAF program in moving passengers and cargo to support peacetime, contingency, and major operations, DOD should develop appropriate policies and procedures for mitigating risks regarding the CRAF program in order to maintain its viability to meet DOD's future needs.[5] DOD agreed that there is a need for comprehensive policy governing all of the CRAF program elements identified in our report; however DOD did not identify what, if any, specific actions it would take in response to our recommendation. However, DOD updated some of its guidance regarding the CRAF program in 2011, and in fiscal year 2012, AMC began a series of studies that may begin to address this recommendation. Further, multiple DOD-sponsored studies have identified risks to the CRAF program resulting from a changed business environment and an expected decrease in CRAF participation—especially in the level of charter airline participation—as DOD's airlift workload declines.[6]

House Report 112-479, accompanying a bill for the National Defense Authorization Act for Fiscal Year 2013, mandated us to report to the congressional defense committees on matters related to the long-term ability of CRAF to meet DOD's needs. [7] Accordingly, we assessed the extent to which DOD has (1) met its military airlift training requirements while also using CRAF participants to the maximum extent practicable, (2) provided justification for restricting commercial carriers from transporting partial plane loads of cargo over certain routes, and (3) established future requirements for the CRAF program, and how the planned size of CRAF compares to those requirements.

To assess the extent to which DOD has met its military airlift training requirements while also using CRAF participants to the maximum extent practicable, we analyzed policies, regulations and guidance pertaining to the management of the CRAF program. We also analyzed DOD's training requirements for its flying hour program to determine the number of hours DOD flew its strategic airlift fleet compared to the number of hours it was required to fly to meet training requirements over the last 10 years.[8] In addition, we conducted interviews with officials from TRANSCOM, AMC, and The Office of the Under Secretary of Defense for Acquisition, Technology and Logistics to understand how DOD decides whether to use the military fleet or commercial airlift. We also interviewed officials from 21 of the 30 CRAF carriers who were participating in the program as of October 2012, as well as representatives from the National Air Carrier Association—an industry association representing several CRAF participants—to understand the perspectives of commercial participants in the CRAF program. To assess the extent to which DOD provided justification for restricting commercial carriers from transporting partial plane loads of cargo over certain routes, we reviewed DOD policies that place restrictions on commercial carriers flying over certain routes. We also reviewed data from fiscal years 2000 and 2012, and interviewed TRANSCOM officials, to obtain information on the rationale and benefits of these restrictions. In addition, we conducted interviews and obtained written responses from CRAF participants to gain an understanding of how such restrictions were affecting their business. To assess whether DOD has established future requirements for the CRAF program and how the planned size of CRAF compares to those requirements, we reviewed studies assessing DOD's strategic airlift capabilities, such as DOD's Mobility Capabilities and Requirements Study–2016, and the Air Mobility Command 2012 CRAF Phase 1 study. We also reviewed data on the number of commercial aircraft committed to DOD's CRAF program and compared these data with DOD's current airlift requirements. Lastly, we interviewed TRANSCOM and AMC officials to determine what steps are being taken to establish future requirements and to gain their perspective on the challenges, if any, they expect to face as they continue to manage the CRAF program.

We conducted this performance audit from August 2012 to June 2013 in accordance with generally accepted government auditing standards. Those standards require that we plan and perform the audit to obtain sufficient, appropriate evidence to provide a reasonable basis for our findings and conclusions based on our audit objectives. We believe that the evidence and data obtained was sufficiently reliable for our purposes, and provides a reasonable basis for our findings and conclusions based on our audit objectives. More detailed information on our scope and methodology is provided in appendix I.

BACKGROUND

The Civil Reserve Air Fleet (CRAF) is a voluntary, contract-based agreement between DOD and U.S. commercial air carriers that augments DOD's military airlift capability during times of war and national emergency. It was created in 1951 to augment DOD airlift capability during a national defense-related crisis. The National Airlift Policy, signed by President Reagan in 1987 and still in effect, establishes policy that the military will rely on the commercial air carrier industry to provide the airlift capability required beyond that available in the military airlift fleet.[9] The policy includes guidelines for meeting airlift requirements in both peacetime and wartime. These guidelines direct, among other things, that policies be designed to increase participation in CRAF and enhance the mobilization base[10] of the U.S. commercial air carrier industry. In exchange for this participation, the government provides commercial carriers the opportunity to fly DOD peacetime missions moving passengers and cargo and also sets aside business for CRAF participants in the General Services Administration City Pairs passenger program and TRANSCOM's Worldwide Express cargo program. CRAF is divided into three progressive stages that TRANSCOM can activate during times of crisis, in part or in whole, with the approval of the Secretary of Defense.

- Stage I covers a minor regional contingency or other situations where AMC cannot simultaneously meet both deployment and other airlift requirements.

- Stage II is tailored for a major theater war or a defense airlift emergency short of a full national emergency.
- Stage III would be required if the military had to fight more than one major theater war at the same time or operate in a larger crisis, including a national emergency declared by the President or Congress. A stage III CRAF activation has never occurred.

DOD has activated CRAF only twice in the history of the program. Stage I and part of stage II were activated in support of Operations Desert Shield and Desert Storm in August 1990 and January 1991, respectively, through May 1991. The CRAF stage I passenger segment was activated in support of Operation Iraqi Freedom in February through June 2003.

To enter the CRAF program, an air carrier must (1) be a U.S. flagged, Federal Aviation Administration approved Part 121 air carrier, (2) be approved by the Commercial Airlift Review Board, [11] (3) have one year prior equivalent uninterrupted service to the commercial sector, (4) meet a minimum fleet participation level (for international carriers), (5) meet a specified utilization rate (for international and aeromedical evacuation fleet participants), and (6) be able to meet manning and crew requirements. Once approved to participate, carriers commit the number of aircraft they will make available for each of the three stages of the CRAF program. AMC then decides the number of aircraft that will be accepted into the CRAF program, based on DOD's wartime requirements. As of April 2013, a total of 64 aircraft were committed to stage I, 308 to stage II, and 554 to stage III.

Two segments of the commercial airlift industry—scheduled service carriers and charter carriers—comprise the CRAF wartime capability. The scheduled service carriers—which include large passenger airlines such as American Airlines and Delta Air Lines and cargo carriers such as FedEx and UPS—pledge the majority of the aircraft accepted into the CRAF program. DOD will use most of the pledged aircraft only during a CRAF activation. In peacetime, scheduled service carriers operate commercial flights on regular routes and cannot afford unplanned disruptions to their airline networks. Because many DOD missions are not routine in their locations or timing, charter carriers—which have the flexibility to provide airlift based on their customers' schedules—transport the majority of DOD's peacetime, contingency, and stage I business.

For some of the charter carriers, this peacetime business accounts for a significant portion of their total business revenue. However, because scheduled service carriers have large fleets, they are also a critical component of CRAF, and they provide the bulk of the CRAF strategic reserve in the event of a CRAF activation. The primary incentive for commercial carriers to participate in the CRAF program is the opportunity to obtain DOD peacetime business. DOD distributes peacetime business to CRAF participants using an entitlement process. CRAF carriers are awarded points based on the number of aircraft they commit to the program, the stage to which these aircraft are assigned, and other considerations as applicable to the individual airline.[12] The amount of peacetime business CRAF participants are entitled to is determined in advance of any missions awarded. DOD makes this business available to the CRAF carriers to fulfill its peacetime business obligation to them, and it does so by offering the carriers the opportunity to fly various missions (for a list of all CRAF carriers, see appendix III).

TRANSCOM and AMC share responsibility with respect to CRAF policy. TRANSCOM validates the requirements for the movement of personnel and cargo, determines which transportation mode will be used for these movements, and distributes the work to the appropriate component command. Once TRANSCOM determines that a movement will go by air, the mission requirement is handled by AMC. Within AMC, the Tanker Airlift Control Center (TACC) normally handles mission planning, assignment of airlift assets, mission control, and tracking. Mission planning includes determining whether military or commercial aircraft will fly a mission. CRAF carriers generally have priority over non-CRAF carriers for movements of passengers and cargo. The Fly CRAF Act generally requires DOD to use CRAF carriers when contracting for airlift services, whenever the carriers are available.[13] If no CRAF participant is available to supply the airlift, DOD may use a non-CRAF carrier (either U.S. or foreign flagged) to fly the mission.[14] For airlift services between two locations outside the United States, CRAF carriers must be used as long as they are "reasonably available." Only foreign carriers operate larger aircraft, such as the AN-124 and IL-76, which are designed to carry outsized and oversized cargo that U.S. commercial carriers normally cannot accommodate.[15] However, according to TRANSCOM officials, DOD uses foreign carriers through subcontracts with CRAF participants, and only rarely contracts directly with foreign carriers. DOD interprets the Fly CRAF Act as applying only to contracts that are specifically for airlift services, and not to contracts for services or supplies that may involve airlift or other transportation services. For example, according to TRANSCOM, DOD does not require "The Fly CRAF preference" to be applied to service or supply contracts such as the Logistics Civil Augmentation Program or the Defense Logistics Agency Prime Vendor Program. According to DOD officials, the current law and related contracting provisions provide the department with the flexibility to acquire the best value for products and services when executing service or supply contracts.[16]

DOD'S PROCESS FOR MONITORING FLYING HOURS PROVIDES INFORMATION THAT COULD HELP IT DETERMINE WHEN TO SHIFT ELIGIBLE MISSIONS TO CRAF PARTICIPANTS

DOD has exceeded the flying hours needed to meet military training requirements for fiscal years 2002 through 2010 due to increased operational requirements associated with Afghanistan and Iraq; however it does not know whether it used CRAF participants to the maximum extent practicable during this period. In fiscal years 2010 through 2012, DOD's flying hours have more closely matched its training plan. In keeping with its policy to both provide training within the military airlift system and use commercial sources of transportation to conduct eligible airlift missions, DOD has taken steps to provide CRAF participants with peacetime business. However, DOD does not use information from its process for monitoring flying hours to determine when it will use more hours than it has planned to meet training requirements and shift eligible airlift missions to CRAF participants to ensure that commercial sources are used to the maximum extent practicable, as required by DOD guidance. Unless DOD uses its information on flying hours to determine when it can shift eligible airlift missions to CRAF participants, it may be flying its military fleet unnecessarily. DOD officials say that using the military fleet to fly missions that are eligible to be shifted to CRAF participants is more expensive than using the CRAF carriers and could reduce these carriers' level of participation in the CRAF program.

DOD Exceeded Its Required Military Training Hours for Fiscal Years 2002 through 2010, but Has Reduced the Percentage of Excess Hours in Recent Years

The National Airlift Policy states that the "Department of Defense shall establish appropriate levels for peacetime cargo airlift augmentation in order to promote the effectiveness of the Civil Reserve Air Fleet and provide training within the military airlift system."[17] Consistent with that policy, DOD Instruction 4500.57 requires that DOD operate its fleet to meet its training requirements and also requires that it use commercial sources of transportation to the "maximum extent practicable."[18] DOD officials stated that they have been using military airlift beyond what was planned, because the operations in Afghanistan and Iraq created additional airlift requirements, many of which could not be met using U.S. commercial sources. For example, some kinds of cargo—such as the mine resistant ambush protected vehicles—are too large to fit inside the aircraft operated by CRAF participants. Military aircraft, along with some foreign aircraft such as the AN-124 and the IL-76, are able to accommodate these kinds of cargo. Additionally, missions in Afghanistan and Iraq often could not be flown by CRAF participants because of airspace restrictions on U.S. carriers operating in those countries. Finally, some missions have additional requirements that call for the use of military airlift, such as requirements that cargo be escorted by military personnel or that an aircraft land on an unpaved runway.

Every year, DOD develops requirements for its military aircrews that serve as the basis for its flying hour program. The flying hour program provides training and experience for the aircrews. These requirements consist mainly of two types of flying hours— "currency hours" and "experiencing hours." Training flights conducted to log currency hours generally do not carry cargo or passengers and therefore do not compete with commercially-operated missions. On the other hand, experiencing, or "time in the air," flights typically carry cargo or passengers and compete with commercially-operated missions. Officials told us that currency hour flights account for roughly 20 percent of the flying hour requirement and are funded through operations and maintenance funds, while experiencing hour flights account for approximately 80 percent of the requirement. We excluded currency hours from our analysis, since flights that provide these training hours generally do not compete for cargo or passengers with commercially-operated missions. However, flights that provide experiencing hour training are funded through the Transportation Working Capital Fund, because these flights carry cargo or passengers. As a revolving fund account,[19] the Transportation Working Capital Fund relies on customer reimbursements in exchange for transportation services provided. The customer that requests airlift reimburses the fund for the mission performed, although some costs associated with mobilization capability and readiness may be funded by the Air Force. For the purposes of this report, "military airlift training requirements" refers to experiencing hours, because those hours are the ones that DOD must decide how to allocate to meet military airlift training requirements while also using CRAF participants to the maximum extent practicable.

Figure 1 shows the percentage by which AMC has exceeded the flying hours that it planned for experiencing requirements in fiscal years 2001 through 2012. DOD said that during these years it exceeded its flying hours for training because of the need to fly missions to support operations in Iraq and Afghanistan. To develop this chart, we compared AMC's requirements for experiencing flying hours to the actual hours AMC flew with the primary airlift platforms—the C-5, C-17, and C-130— and expressed them as percentages of the planned flying hours. We excluded the tanker aircraft from this analysis, since there are no commercial aircraft in the CRAF program that are comparable to the KC-10 or KC-135.

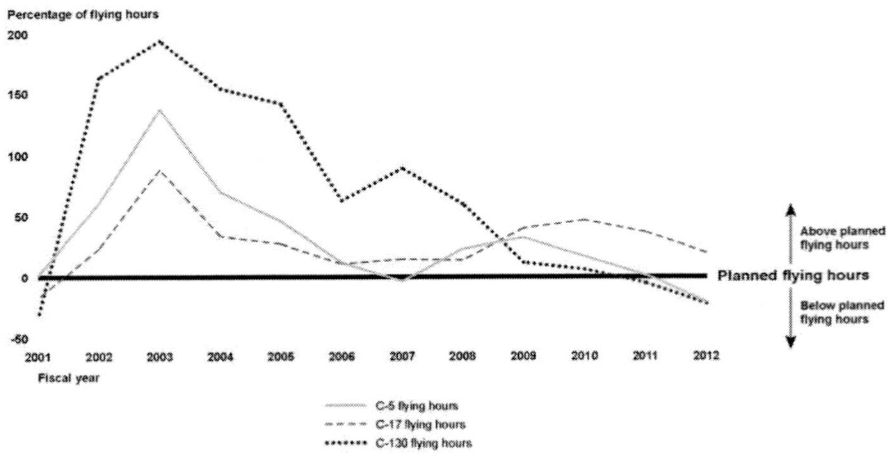

Source: GAO analysis of AMC data.

Note: The flying hours in this chart include only Transportation Working Capital Fund hours that are used for experiencing training and airlift movements. DOD balances these airlift movements between commercial and military aircraft.

Figure 1. Percentage by Which Actual AMC Flying Hours Exceeded Planned AMC Flying Hours in Fiscal Years 2001–2012.

DOD Has Taken Steps to Provide CRAF Participants with Peacetime Business

Recognizing the importance of the commercial carriers for meeting its future airlift requirements, DOD has taken steps to increase the amount of peacetime business it gives to CRAF participants. According to TRANSCOM's Fiscal Year 2012 Annual Report,[20] CRAF carriers remain essential in supplying transportation services and provide a critical part of DOD's war fighting deployment capability. Further, TRANSCOM and AMC are using CRAF carriers to more directly support the forces in Afghanistan. CRAF participants have provided the majority of passenger movements and about a quarter or more of all cargo movements since fiscal year 2004. Figures 2, 3, and 4 show the extent to which DOD has relied on CRAF participants to provide airlift services.

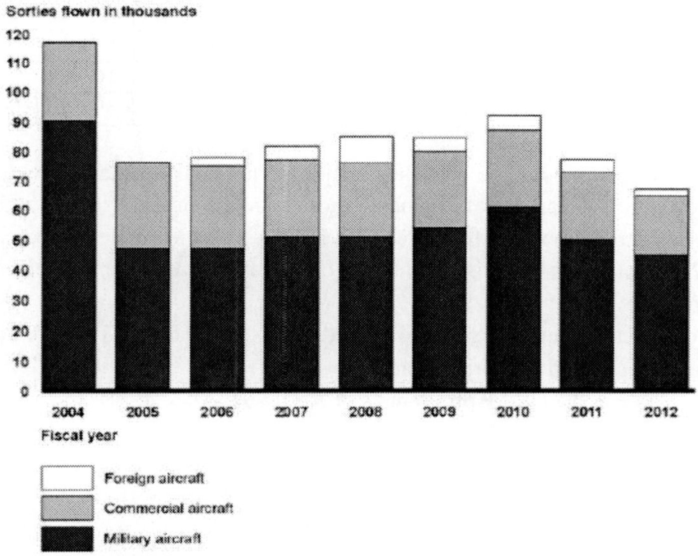

Source: GAO analysis of AMC data.
Note: A sortie is defined as the takeoff and landing of one aircraft.

Figure 2. AMC-Managed Sorties Flown in Fiscal Years 2004–2012, by Airlift Source.

Over the last few years, both the number of CRAF participants and the number of aircraft pledged to the CRAF program have fluctuated, and it is not clear what level of support CRAF participants will provide in the future. For example, as we noted in our 2009 report on CRAF,[21] the number of charter aircraft enrolled in the CRAF program had declined from more than 60 aircraft in 2003 to as few as 19 in April 2008, before stabilizing at 29 charter aircraft in May 2008. Our analysis shows that CRAF participation as of fiscal year 2012 was still sufficient to allow DOD to meet its wartime requirements. However, according to some current CRAF participants, changes to the business environment, such as the ongoing economic down-turn, have resulted in five of the participating carriers filing for bankruptcy over the last three years. Two of these carriers have already completely ceased providing airlift services. Table 1 shows the level of airlift support provided by CRAF participants and military aircraft during the last three years.

Table 1. Level of Airlift Support Pledged by CRAF Participants for Stage 3 in Fiscal Years 2011–2013

Civil Reserve Air Fleet (CRAF) Support		2011	2012	2013[b]
Participating Carriers		35	31	28
Number and Type of Aircraft Pledged to CRAF	Passenger	686	823	392
	Cargo	175	163	162
	Aeromedical[a]	39	39	0

Source: GAO analysis of Air Mobility Command data.

Note: Aircraft noted above are included in the International Segment (Long Range & Short Range), Aeromedical Segment, and National Segment (including Domestic and Alaskan sections). In fiscal year 2011, TRANSCOM implemented changes to the CRAF program limiting the number of aircraft that would be accepted into the program.

[a]The Aeromedical requirement was reduced in fiscal year 2013 and currently no carriers are pledged to this segment.

[b]Aircraft enrolled in the CRAF program as of April 2013.

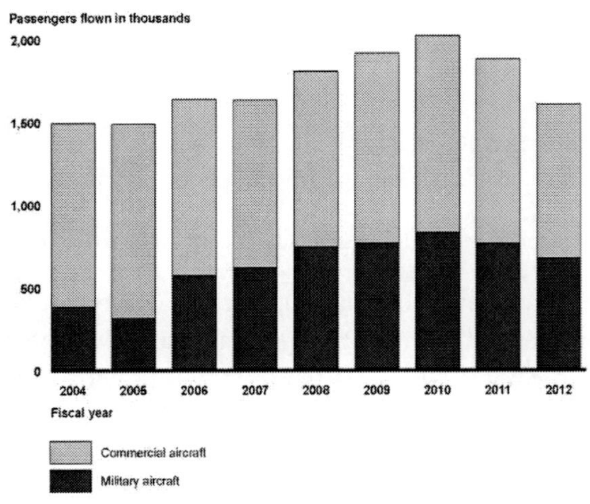

Source: GAO analysis of AMC data.

Figure 3. Passengers Flown on AMC-Managed Flights in Fiscal Years 2004–2012, by Airlift Source.

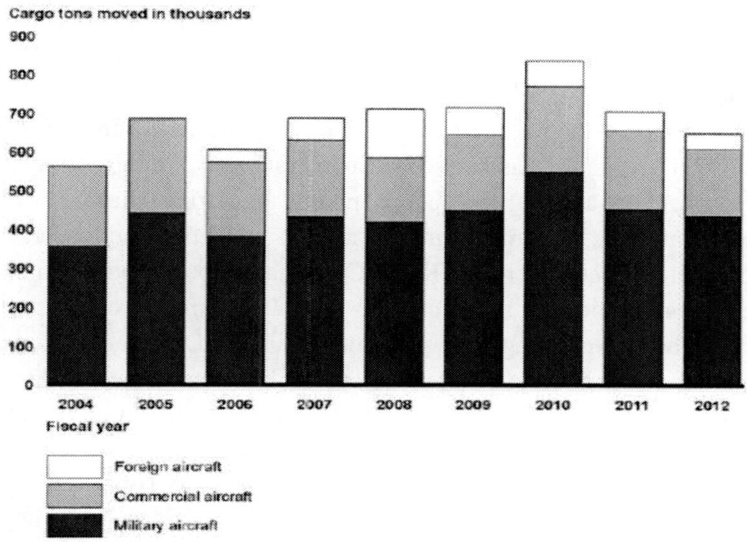

Source: GAO analysis of AMC data.

Figure 4. Cargo Tons Flown on AMC-Managed Flights in Fiscal Years 2004–2012, by Airlift Source.

To support increasing the amount of business provided to CRAF participants, TRANSCOM has created a new organization called the Enterprise Readiness Center. According to an official with the Enterprise Readiness Center, one of the goals of the center is to explore ways to encourage DOD organizations, like the Defense Logistics Agency, to direct more air cargo business into the DOD-managed Defense Transportation System. Further, the center will also seek to preserve DOD's airlift readiness capability, given the reduction in airlift volume, and to help DOD maintain efficiencies by ensuring that the Defense Transportation System is the primary source used by DOD entities to arrange transportation. To achieve this, the Enterprise Readiness Center proposes to improve the usage process of the Defense Transportation System, create flexible rates, minimize overhead as a way to reduce rates, develop customer-based transportation solutions, and create an active dialogue with CRAF participants. As a way to further strengthen communications and the strategic relationship between DOD, the Department of Transportation, and CRAF participants, TRANSCOM and AMC also established an Executive Working Group in 2010. The Executive Working Group is a discussion forum used to address general issues related to the CRAF program. The working group's meetings are a forum for providing updates regarding efforts related to the CRAF program, such as updates on various studies, the status of efforts related to CRAF, and carrier concerns.

DOD officials also told us that they have taken additional steps over the last few years to improve the distribution of business within the CRAF program. TRANSCOM has revised its process for awarding points over the last few years to award more bonus points to carriers that fly additional peacetime missions, assume a greater risk of activation, and operate more modern, fuel-efficient aircraft. TRANSCOM has also revised the practice of awarding commissions. Larger carriers allow the smaller carriers on their teams[22] the benefit of using their points to obtain DOD business, in exchange for commissions consisting of a percentage of the revenue the smaller carriers earn from this business. These commissions are one of the ways in which larger carriers earn revenue from the CRAF program, since they do not conduct many of the actual airlift missions in peacetime. However, according to an official at one carrier, these commissions had risen to as high as 9 percent of the revenue earned from the mission. TRANSCOM officials told us that they have capped the value of these commissions at 5 percent of mission revenue, in an attempt to ensure that smaller carriers earn enough profit from performing peacetime airlift missions.

DOD intended for these efforts to strengthen the viability of the CRAF program. The opinions of CRAF participants varied on the extent to which these changes made the program more equitable, mostly depending on whether the carrier directly benefited from the changes. All of the carriers we spoke with indicated that they were planning to stay in the CRAF program for the immediate future. However, some added that if the revenue they were receiving decreased too much, they would reassess their participation and would consider not participating in future years. More than half of the CRAF participants we interviewed suggested that DOD could do more to increase the peacetime business it provides to them. Some of these carriers suggested that DOD's use of foreign air carriers should be curtailed. According to DOD officials, foreign carriers primarily operate as sub-contractors to CRAF participants to move cargo that is too large for standard U.S. commercial aircraft, and only in rare cases would DOD contract directly with a foreign carrier. Furthermore, our analysis indicates that the use of foreign carriers has declined since its high point in fiscal year 2008. As shown in figure 5, payments made to foreign carriers have declined by more than 55 percent since fiscal year 2008.

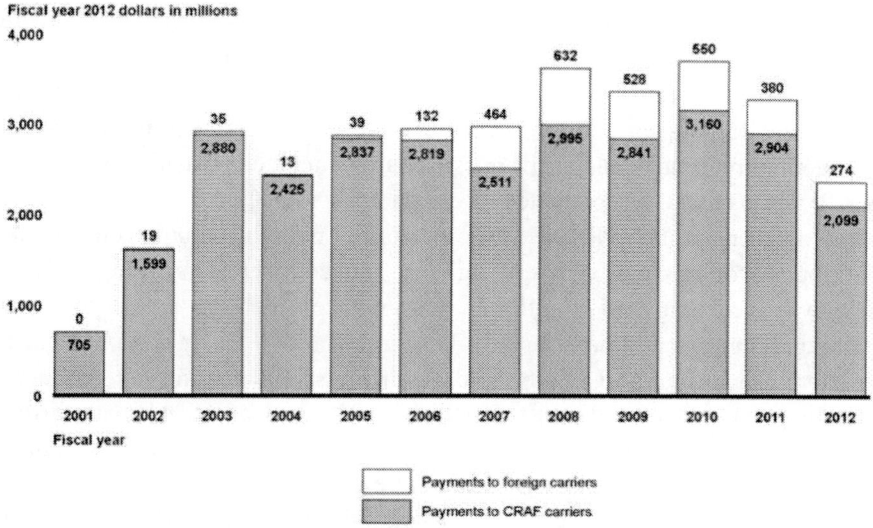

Source: GAO analysis of TRANSCOM data.

Figure 5. DOD Payments Made to CRAF Participants and Foreign Carriers in Fiscal Years 2001–2012.

DOD Does Not Use Its Process for Monitoring Training Hours to Inform Its Allocation of Eligible Airlift Missions to CRAF Participants

DOD does not use its process for monitoring flying hours to determine when it will exceed its planned training hours, and it does not use the information from this process to allocate eligible airlift missions to CRAF participants. As previously noted, DOD guidance requires TRANSCOM to meet its training needs while also using commercial sources of transportation to the "maximum extent practicable." DOD officials told us that, consistent with this policy, meeting training needs was their priority. However, they also told us that flights provided by CRAF participants are less expensive than military flights, in part because commercial aircraft are designed to be more fuel-efficient, while military aircraft are designed to carry heavy cargo and land in austere locations. In addition, according to AMC data, once training requirements have been met, using commercial carriers for airlift missions can be less costly than using military aircraft. For example, according to an April 2013 analysis provided by AMC officials, the cost per pound to transport cargo using commercial carriers such as the 747 and MD-11 can be between 22 and 35 percent lower than the cost of transporting the same cargo using military aircraft such as the C-5 and C-17. Currently, airlift requests are handled by different sections within the Tanker Airlift Command Center (TACC), depending on the type of airlift requested. Each of these sections has a different process for choosing whether to use commercial or military airlift to meet the request. Some airlift missions are conducted primarily by military airlift, while others are conducted by commercial sources.

However, while TRANSCOM performs periodic monitoring of the distribution of missions between military and commercial sources, officials acknowledged that this monitoring does not consider the extent to which training requirements have already been met or will be met with planned missions. According to DOD officials, airlift missions that are not conducted to satisfy training requirements should be performed by CRAF participants, except when there is some other feature of the mission that requires military airlift. Knowing when more flying hours are going to be used than are needed to meet training requirements—and using this information to shift eligible airlift missions to CRAF participants—would allow DOD to use commercial sources of transportation to the maximum extent practicable.

DOD officials told us that operations in Iraq and Afghanistan had ensured that there were enough airlift missions available both to support training requirements and to provide adequate peacetime business for CRAF participants. Further, they noted that there are a number of reasons that DOD might exceed its flying hours, such as the need to transport particularly large cargo, special conditions that require military aircraft (such as unpaved runways), and restrictions on U.S. carriers operating in Iraq and most of Afghanistan. Given such requirements, officials questioned the utility of developing a process to monitor the balance between satisfying flying hour training requirements and providing CRAF participants with additional peacetime business; they said that they were uncertain how many additional missions would be eligible to be flown by commercial carriers. However, TRANSCOM and AMC officials have acknowledged that they have not collected data that would allow them to determine how many of these missions could be shifted to CRAF participants. Furthermore, while we acknowledge that there may be a number of legitimate reasons why military aircraft would have to be used for missions even after training requirements have already been met, it is not clear that such reasons are always present when military airlift is used. For example, AMC completed a study in December that was intended, in part, to address short-term concerns regarding the CRAF program and its participants. This study noted that some missions were flown on military aircraft only because the necessary load plans[23] for commercial aircraft had not been developed in a timely manner—not because of any requirement that the cargo be flown on military aircraft. The study recommended that DOD airlift customers develop commercial load plans to facilitate scheduling of commercial aircraft in these situations. This study acknowledges that some missions currently flown by military airlift could instead be flown by CRAF participants without negatively affecting training hours.

After the drawdown in Afghanistan concludes, the need for airlift is expected to decline, which will reduce both training opportunities and the business available for CRAF participants. In addition, as airlift needs decrease, DOD may need to fly a higher percentage of its channel missions in order to provide its crews with sufficient training opportunities, which could further decrease its use of CRAF participants. DOD officials told us that they expect peacetime business to fall significantly after fiscal year 2015. This decrease has already begun; peacetime revenues of CRAF participants have already dropped by nearly one third, from their high point of approximately $3 billion in fiscal year 2010 to about $2 billion in fiscal year 2012, as shown in figure 5. Commercial carriers are projected to be used even less in fiscal year 2013 and beyond, until revenues return to pre-September 11, 2001 levels of $700 million or less. This represents a potential 66 percent decline in DOD business available to CRAF participants, which may further exacerbate the economic pressures under which CRAF participants are operating.

By not using the information it has on flying hours to help determine when it can allocate eligible airlift missions to CRAF participants, DOD loses the ability to determine whether it is using commercial sources—such as CRAF participants—to the maximum extent practicable, as required by DOD guidance. As a result, DOD may be using its military fleet more than necessary thereby risking reduced participation of commercial carriers in the CRAF program.

DOD Restricts Partial Plane Loads on Channel Routes to Promote Efficiency, Meet Training Requirements, and Fulfill Peacetime Business Obligations to CRAF

DOD provided several reasons for restricting commercial carriers from transporting partial plane loads of cargo over certain routes, based on its need to promote efficiency, meet its military airlift training requirements, and fulfill peacetime business obligations to CRAF participants. According to TRANSCOM officials, in 2001, DOD began restricting commercial air carriers from transporting partial plane loads of cargo over certain overseas channel routes in order to improve the efficiency and effectiveness of the cargo missions flown over these routes and keep cargo flights in the channel route system that DOD relies on to satisfy its training requirements and business obligations to CRAF participants. In May 2012, TRANSCOM issued a memorandum reiterating its policy of restricting commercial aircraft—including CRAF participants—from transporting partial plane loads of cargo over these routes.[24] According to TRANSCOM officials responsible for coordinating airlift for DOD, this policy—which has been in place for over a decade—is a tool to help DOD increase the efficiency of its cargo shipments airlifted over channel routes and minimize costs to DOD. DOD officials reported that in the late 1990's and early 2000's commercial air carriers began transporting an increasingly larger share of DOD cargo shipments, leaving a relatively small amount of cargo for military aircraft to transport over channel routes. DOD officials said that before the policy was implemented, military aircraft would often conduct channel route missions with partial loads of cargo, instead of completely filling the aircraft with cargo, which is more cost effective. In addition, in the late 1990's and early 2000's, DOD was experiencing a shortage of flying hours for training. During this same period, commercial carriers were flying a large number of airlift missions, which exacerbated DOD's flying hour shortage, because many of the airlift missions that military aircrews could have conducted for training purposes were being lost to commercial air carriers. Lastly, according to TRANSCOM officials, because many of the partial planeload missions performed by commercial carriers were negotiated under tender contracting arrangements— which are not included in the annual amount of peacetime business DOD guarantees to the CRAF program—DOD's ability to fulfill its peacetime business obligations to CRAF was being challenged.

The National Airlift Policy states that military and commercial resources are equally important— and interdependent—in fulfilling the national defense airlift objective.[25] The policy also provides that the goal of the U.S. government is to maintain in peacetime military airlift resources that are manned, equipped, trained, and operated to ensure the capability to meet wartime requirements. DOD guidance also notes that TRANSCOM may be required to maintain a readiness posture that includes operating military airlift internationally during peacetime, and that it must conduct such operations at the level necessary to meet operational and training requirements.[26] According to DOD officials who are responsible for managing DOD's strategic airlift requirements, TRANSCOM takes steps to meet DOD's flying hour training requirements while also providing commercial carriers with peacetime business; however the flying hour training requirement takes precedence. DOD performs a variety of types of airlift missions that allow military aircrews to meet their flying hour training requirements while also delivering the cargo needed to sustain military operations to military units located overseas. These mission types include:

- **Channel airlift missions:** regularly scheduled airlift for movement of sustainment cargo and/or personnel between designated aerial port of embarkation and aerial port of debarkation over validated contingency or distribution channel routes.
- **Special assignment airlift missions:** airlift missions requiring special pickup/delivery at locations other than those established within the approved channel structure or requiring special consideration because of the number of passengers, weight or size of the cargo, urgency or sensitivity of movement, or other special factors.
- **Contingency missions:** airlift for movement of cargo and/or personnel in support of military operations directed by appropriate authority to protect U.S. interests.
- **Exercise missions:** airlift for movement of cargo and/or personnel in support of a military maneuver or simulated wartime operation involving planning, preparation, and execution that are carried out for the purpose of training and evaluation.
- **Theater direct delivery:** a theater-based distribution system wherein delivery to destinations forward of major aerial ports of debarkation can be performed by any available aircraft, including those normally used for intertheater requirements.

Channel route missions are conducted by both military and CRAF participants and account for a large portion of DOD's overall airlift activity. During the last three fiscal years, at least 30 percent of DOD's total cargo movement was over channel routes. See figure 6 below.

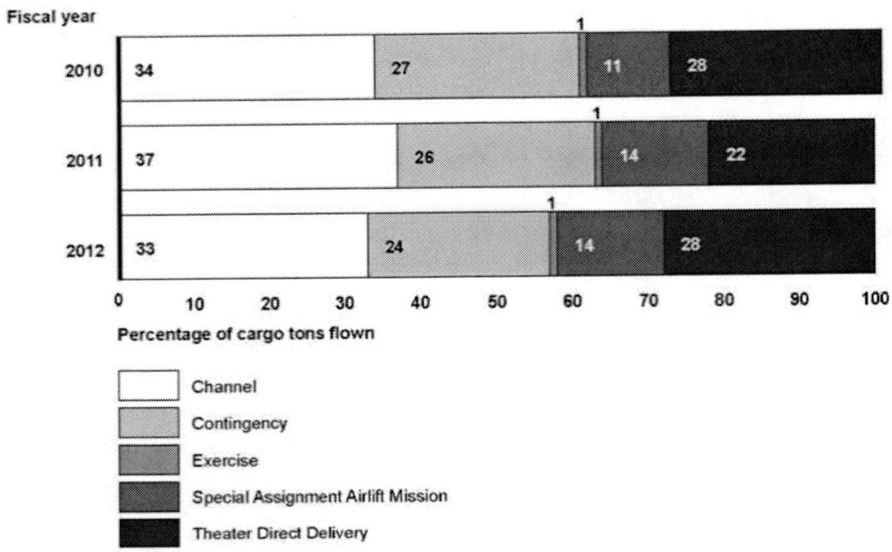

Source: GAO analysis of AMC data.

Note: Percentages may not add up to 100 percent due to rounding.

Figure 6. Fiscal Years 2010–2012 Cargo Tons Flown by Airlift Mission Type (as Percentages of All Airlift).

TRANSCOM officials stated that to maximize efficiency, DOD requires aircraft conducting channel route missions—whether they are military or commercial—to be completely full of cargo before takeoff. According to TRANSCOM officials, the policy restricting commercial carriers from transporting partial loads over channel routes provides DOD with a tool to maximize the amount of cargo transported in a single mission over a channel route. Cargo previously transported by commercial carriers in partial loads is now consolidated at aerial ports of embarkation. TRANSCOM officials reported that historically, commercial carriers transporting partial loads had been conducting a large portion of DOD's airlift business. These commercial airlift missions involved transporting cargo to and from locations that were also being serviced by military aircraft conducting channel missions. DOD was not maximizing the efficiency of its channel route missions and minimizing costs, because aircraft were not filled to capacity. To reduce the redundancy of transporting cargo using both modes of delivery, DOD began restricting commercial carriers from conducting partial plane load missions over channel routes, and it now generally requires commercial aircraft conducting channel missions to be full of cargo before takeoff. According to

TRANSCOM officials, the policy ultimately played a role in increasing the efficiency of DOD's air cargo movements over channel routes. According to a RAND report issued in 2003 that analyzed the costs associated with transporting cargo over channel routes using commercial airlift versus military airlift, DOD would decrease airlift cost if it reduced the amount of cargo transported by commercial carriers conducing partial plane load missions, and shifted that cargo to be transported in a full plane load aircraft.[27] Taking this step would be less expensive than allowing military aircraft to conduct partial plane load missions over channel routes.

In addition, DOD's policy allows it to offer more training opportunities for its aircrews during periods of low demand for airlift. Rather than relying on other types of missions—such as contingency missions— to accomplish training, AMC prefers to schedule flying hours for training on channel route missions, which are regularly scheduled, planned in advance, consistent, and predictable. Channel route missions are used to maintain and upgrade pilots' flying skills and, as part of the training, can include transporting cargo from specific military locations within the United States—such as McGuire Air Force Base in New Jersey—to overseas military bases located in countries like Germany or Kuwait. These missions are conducted on a regularly scheduled basis and include DOD cargo, so they provide commanders with reassurance that they will receive planned amounts of sustainment cargo within a designated time frame.

TRANSCOM officials told us that in the late 1990's and early 2000's commercial aircraft had been conducting a large portion of DOD's airlift business, but the overall demand for DOD airlift was relatively low; as a result the military began experiencing a shortage of flying hours to use for training. Many of the airlift missions flown by commercial carriers involved transporting cargo that could have been transported by military aircraft. DOD's policy of restricting commercial carriers from transporting partial loads over channel routes has allowed DOD to shift cargo into the channel route system, increase the number of channel route missions available for aircrews to satisfy flying hour training requirements, and address DOD's flying hour shortage. In 2003, the RAND Corporation conducted a study[28] of the peacetime tempo of DOD's air mobility operations and asserted that DOD needed to take steps to address its shortage of flying hours. The report found that, during fiscal year 2000 and 2001, aircrew personnel encountered a flying hour shortage because international military activity was relatively calm and there were fewer U.S. missions that required airlift support. The report also pointed out that because commercial carriers had begun conducting a large portion of DOD's airlift business, decreasing the amount of airlift business given to commercial carriers would help reverse this trend and help alleviate DOD's flying hour shortage. The report's conclusions supported measures taken by DOD to implement a policy to decrease peacetime business provided to commercial carriers when necessary, to support training requirements. With DOD's policy in place, more cargo was being funneled into the channel route system, and DOD was able to increase the number of channel route missions offered to military aircrews, thereby helping to alleviate the shortage of flying hours.

Further, TRANSCOM officials said that DOD's policy of restricting commercial carriers from transporting partial planeloads of cargo over certain channel routes was also implemented in part to help DOD fulfill its peacetime business obligations to CRAF. Through the CRAF peacetime airlift contract, DOD provides a certain level of airlift business to CRAF participants; DOD negotiates a designated amount of business that it is committed to provide to CRAF participants—as an incentive for commercial carriers to participate in the CRAF program—and distributes this business among the CRAF participants currently enrolled in the program. This business consists, in part, of missions flown across channel routes. TRANSCOM officials reported that many airlift missions conducted by commercial carriers carrying partial loads across channel routes were being arranged through tender-based contractual agreements. Tender-based agreements for airlift services are offers by an air carrier to provide transportation services at a specified rate. According to TRANSCOM officials, business associated with tender- based contracts falls outside of the CRAF peacetime business entitlement obligation. TRANSCOM officials said that this practice was diminishing the pool of peacetime business that DOD could provide to CRAF participants under the CRAF peacetime business entitlement process.

According to TRANSCOM officials, the policy limiting the amount of tender-related airlift business provided to commercial carriers increases the efficiency of channel route missions, alleviates the shortage of flying hours for training, and allows DOD to provide CRAF participants with peacetime business to fulfill its CRAF peacetime business obligations. In addition, TRANSCOM officials said that in periods of high demand for airlift, such as the last several years, DOD can provide CRAF participants with more channel route business, because military aircrews can satisfy their training requirements for flying hours by conducting other airlift missions, such as contingency and special assignment airlift missions.

Some CRAF participants expressed concerns to us that the original rationale for DOD's policy no longer exists and that the policy may prevent DOD from using the less costly commercial airlift option to transport partial loads of cargo over channel routes. First, according to TRANSCOM officials, the original rationale for the policy was to ensure that DOD could provide sufficient flying hours to train its aircrews. Two of the CRAF participants we interviewed stated that this policy was no longer necessary because DOD no longer faces flying hour shortages as it did in the late 1990's and early 2000's. DOD officials stated that it is important to retain this policy as a management tool, especially since DOD's need for airlift is projected to return to pre-September 2001 levels by 2015. According to DOD officials, data from fiscal year 2000 illustrate this point; in 2000, DOD needed to reserve about 57 percent of its channel route missions for training and provided about 28 percent of the channel route missions to CRAF participants. In contrast, during a period of high demand for airlift, such as fiscal year 2012, DOD reserved about 31 percent of its channel route missions for training and was able to provide more than 60 percent of its channel route missions to CRAF participants. See figure 7 below.

Second, a CRAF participant we met with emphasized that using commercial aircraft to transport partial loads is less costly than using military aircraft to transport partial loads, because when using commercial airlift, DOD pays by the pound and only for the cargo airlifted, rather than incurring the entire cost of using a military aircraft to carry a partial load. DOD officials acknowledge that using commercial carriers to transport partial plane loads of cargo is less expensive than using military aircraft for this purpose and note that the policy restricting commercial carriers from transporting partial plane loads of cargo over certain overseas channel routes has a provision to allow commercial carriers to conduct such missions on a case-by-case basis, when needed to meet DOD's requirements. For example, if a customer requires a critical, time-sensitive item and cannot wait for it to be transported by a regularly scheduled channel mission, commanders still may have the option to use a commercial carrier to transport a partial load to a designated location using a channel route.

In addition, commercial carriers can transport cargo outside the channel route system under a variety of other DOD airlift transportation contracts. For example, TRANSCOM's World-Wide Express program, an airlift transportation program available only to CRAF participants, is used to provide international commercial express package transportation service for shipments up to and including 300 pounds. This program provides DOD with the ability to ensure that commanders can receive unique, time-sensitive cargo items when no channel mission is available within a specified time frame. Over the last five years, this program has consistently generated over $100 million dollars worth of airlift business annually for CRAF participants.

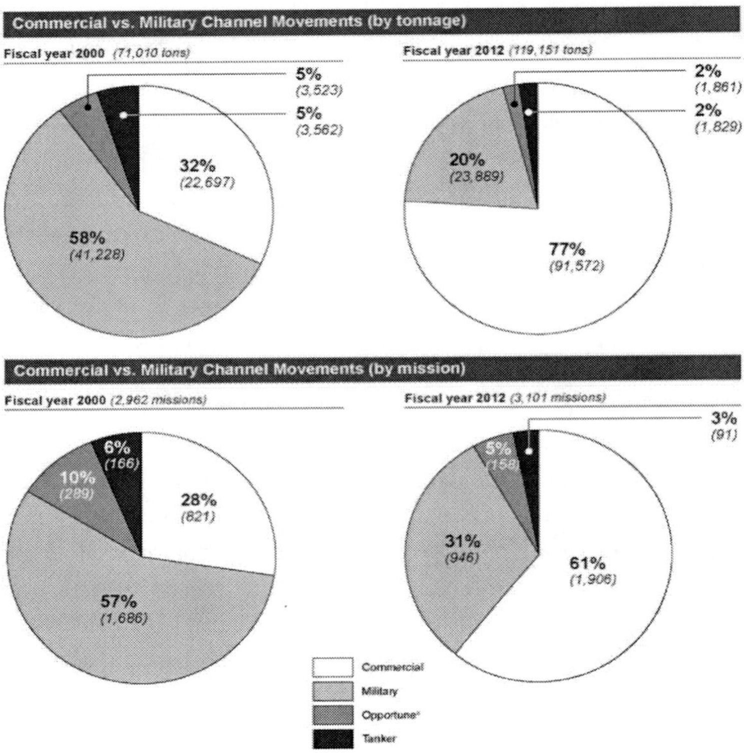

Source: GAO analysis of AMC's FY 2000 and FY 2012 Channel Performance and Analysis Reports.

Note: Percentages may not add up to 100 percent due to rounding.

[a]Opportune missions are specific training missions not already committed to a designated airlift mission, such as contingency or exercise airlift missions. Military aircraft conducting opportune missions transport cargo in the remaining space available aboard the aircraft.

Figure 7. Commercial versus Military Channel Route Movements for Fiscal Years 2000 and 2012.

According to TRANSCOM and AMC analysis, as the drawdown efforts in Afghanistan proceed over the next few years, airlift demand is expected to decline to pre-September 11, 2001 levels. It will therefore be important for

DOD to plan and ensure that military aircrews are provided with ample opportunity to fulfill training requirements and avoid a shortage of flying hours. In preparation for this decline in the demand for airlift, TRANSCOM officials emphasized that DOD's policy to restrict commercial carriers from transporting partial loads over channel routes may continue to serve as an important management tool and allow DOD to balance the goals of operating its channel route system as efficiently as possible, providing enough training opportunities to military aircrews, and fulfilling its CRAF peacetime business obligations.

Adequacy of CRAF to Meet Future Mission Needs is Unclear until DOD Completes Several Assessments

DOD is conducting several interrelated studies to determine its future airlift requirements; however it is unclear whether the planned size of CRAF will be adequate to meet future airlift requirements. The National Defense Authorization Act for Fiscal Year 2013 requires DOD to conduct a study that assesses its mobility needs—referred to as the Mobility Requirements and Capabilities Study 2018,[29] which DOD had not begun at the time of our review. In addition, in response to the changing business environment, AMC is also conducting a two-phase study to assess the readiness of CRAF participants to augment DOD's airlift capacity and the viability of the CRAF program. The CRAF Phase 1 study was completed in December 2012, and according to officials the Phase 2 is scheduled to be completed in the fall of 2013. Meanwhile, DOD has been taking steps to continue to encourage commercial carriers to participate in the program. Until DOD finalizes these assessments, it will be unclear whether the planned size of CRAF will be adequate to meet future airlift requirements.

DOD's Mission Needs for CRAF Have Changed Since its Last Assessment of Future Requirements

DOD reports that there are more aircraft committed to the CRAF program than are needed to fulfill the wartime requirements established by the Mobility Capability Requirements Study 2016 (MCRS–16), which was issued in 2010. However, it is not clear whether the current level of CRAF participation will provide the right number and mix of aircraft to meet future requirements, since

DOD has issued new strategic guidance that may affect DOD's airlift requirements. While the number of aircraft pledged to the program has fluctuated, DOD's past analysis showed that the projected size and mix of the CRAF fleet was more than adequate to satisfy all war planning scenarios established by the MCRS–16. According to DOD data, as of March 2012, CRAF participants had enrolled 15 percent more aircraft in the program than would be needed to meet established airlift requirements. The MCRS–16 assessed components of the mobility capabilities that DOD would need for possible future strategic environments and was intended to help DOD make investment decisions regarding mobility systems, such as how much to invest in strategic airlift to meet wartime needs. Among other things, the study examined how changes in the mobility system affect the outcomes of major operations and assessed the associated risks. The MCRS–16 determined that, with few exceptions, the mobility capabilities projected for 2016 would be sufficient to support the most demanding projected requirements. The study assessed the major mobility systems required to move personnel and supplies from their point of origin to their destination: sealift, surface transportation, and airlift components, to include strategic airlift, aerial refueling, and CRAF passenger and cargo. To support decisions regarding future mobility force structure, the MCRS–16 developed three demanding cases consisting of conflicts and natural disasters with multiple scenarios occurring over a 7- year period and requiring the use of mobility capabilities. The MCRS–16 used approved DOD planning scenarios to develop the three cases. For example, in one case, U.S. forces might be required to conduct a large land campaign and a long-term irregular warfare campaign while also responding to homeland defense missions. In another case, U.S. forces might be conducting two nearly simultaneous large-scale campaigns, while also responding to three nearly simultaneous domestic events and conducting other operations.

Since its last assessment of its airlift requirements in 2010, DOD has issued new strategic guidance. Specifically, DOD's strategic guidance issued in January 2012 calls for, among other things, an increased focus on the Asia-Pacific region and resizing U.S. forces,[30] both of which may affect airlift needs. For example, an increased focus on the Asia-Pacific region could affect operational plans in that theater and require changes to the number and type of forces assigned to the region, as well as the associated airlift requirements. In addition, the resizing of DOD forces to achieve security objectives could have implications for the choice of commercial and military aircraft used to support future military operations. In March 2013, the Secretary of Defense tasked DOD senior leadership to examine the department's strategic assumptions, following up on the January 2012 Defense Strategic Guidance which, among other things, called for rebalancing military forces toward the Asia-Pacific region. This review examines the choices underlying the department's strategy, force posture, investments, and institutional management, as well as past and future assumptions, systems, and practices. The results of the review will frame the secretary's guidance for the fiscal 2015 budget and will be the foundation for the Quadrennial Defense Review expected to be issued in February 2014.

DOD Is Updating Its Future Airlift Requirements to Reflect Strategy Changes and Reassess the Incentives for CRAF Participants

The National Defense Authorization Act (NDAA) for Fiscal Year 2013 requires DOD to conduct a new mobility capabilities requirements study—referred to as the Mobility Requirements and Capabilities Study 2018 (MRCS–18)—based in part on the new defense strategy mentioned above.[31] This new assessment may provide decision makers with the analytical data needed to determine DOD's airlift capability requirements and the number and type of aircraft CRAF participants would need to pledge to the program in order to support these requirements. Among other things, the NDAA requires DOD to

- describe and analyze the assumptions made by the Commander of the U.S. Transportation Command with respect to aircraft usage rates, aircraft mission availability rates, aircraft mission capability rates, aircrew ratios, aircrew production, and aircrew readiness rates;
- assess the requirements and capabilities for major combat operations, lesser contingency operations as specified in the Baseline Security Posture of the Department of Defense, homeland defense, defense support to civilian authorities, other strategic missions related to national missions, global strike, the strategic nuclear mission, and direct support and time-sensitive airlift missions of the military departments; and
- identify mobility capability gaps, shortfalls, overlaps, or excesses and assess the risks associated with the ability to conduct operations and recommended mitigation strategies where possible.[32]

Until DOD completes the MRCS–18, decision makers in DOD and Congress may not have all of the relevant information they need to ensure that DOD's mobility capabilities and requirements are sized most effectively and efficiently to support the U.S. defense strategy. DOD acknowledges the requirements set forth in the National Defense Authorization Act for Fiscal Year 2013 and fully intends to cooperate and work to complete the assessment, but according to AMC and TRANSCOM officials, no time frame has been established for when this study will be completed.

Further, AMC has begun conducting additional studies to assess its airlift requirements and how the CRAF program will support near-term and long-term requirements. AMC's CRAF study is being conducted in two phases and will help AMC to ensure that the commercial airlift forces associated with the CRAF program are prepared to support the drawdown of forces in Afghanistan by the end of calendar year 2014. Phase 1 of the CRAF study, completed in December 2012, focused on the international long-range segment of CRAF, which will be most affected by the decreasing demand for airlift resulting from the drawdown of forces in Afghanistan. It identifies a series of issues facing CRAF during the withdrawal and for the short term following the drawdown. A number of observations are directly related to the drawdown period and the period immediately following. These require near-term actions to ensure that commercial airlift support will be available when it is needed to support national interests. For example, the Phase 1 study noted that a future study should assess the risk and reward factors that may affect further CRAF participation due in part to the state of flux in the current charter air industry resulting from economic pressures brought on by a decline in commercial passenger charter opportunities.

In addition to discussing certain recommendations from the Phase 1 study, Phase 2 of the CRAF study will focus on maintaining the future viability of the CRAF program and its readiness to augment military airlift capability and support surge requirements. This follow-on study will undertake an in-depth analysis of issues identified in Phase 1 that could affect the long-term viability and reliability of the CRAF program. The findings from the Phase 2 study will propose courses of action and mitigation strategies to ensure CRAF readiness now and in the future, balancing government interests and mandates with the dynamics of the changing industry. Furthermore, the CRAF Phase 2 study will evaluate the market, the carriers and their business base, and the existing business models within industry and government in order to provide insights and recommend actions to ensure that the CRAF program can continue to meet wartime requirements in the future. AMC and TRANSCOM expect this study to be completed by the fall of 2013.

According to AMC officials, one of the issues that will be addressed in the Phase 2 study is the recommendation from the Phase 1 study that DOD continue the suspension of the 60/40 rule through fiscal year 2014. The 60/40 rule was created as a safeguard for DOD. Under the 60/40 rule, DOD business cannot provide more than 40 percent of a carrier's revenue and the remaining 60 percent of the carrier's revenue must be earned through sources other than DOD, generally referred to as commercial sources or commercial air transportation. Carriers that earn more than 40 percent of their revenue from DOD may be penalized by reductions in their entitlement to DOD business. Prior to fiscal year 2010, the rule was based on an air carrier's revenue. However, in 2010 the rule was modified so that it calculated the percentage of business in block hours[33] rather than amount of revenue. As of May 2010, the rule has been suspended. One of the original goals of the 60/40 rule was to ensure that CRAF carriers maintained a strong commercial business base, efficient operations, and modern fleets to help prevent them from going out of business when DOD demands were low. Limiting the proportion of DOD business carriers could have would also provide DOD with a surge capability to draw on if demand grew suddenly. According to TRANSCOM and AMC officials, the 60/40 rule was suspended so that commercial carriers would not be penalized for supporting increased DOD airlift demands. If carriers continued to increase support while still being required to observe the 60/40 rule, the rule would prove to be counterproductive. DOD would be asking for increased support while potentially issuing penalties to those carriers providing the increased support. Some of the carriers we spoke with stated that the 60/40 rule had not been strictly enforced and that the suspension of the rule had no effect on the amount of business they received as a result of participating in the CRAF program. However, according to DOD officials, five carriers have gone bankrupt in the last three years and two of them have stopped offering airlift services even though this rule has been suspended.

Based on data included in the MCRS–16, DOD counts on the CRAF program to provide most of the passenger airlift services as well as a significant amount of the cargo services to support wartime requirements. Therefore, CRAF must maintain the ability to respond in order to meet combatant commander requirements. DOD must develop accurate requirements if CRAF is to maintain the ability to respond to these requirements. For that reason, until DOD completes the MRCS–18 and the

CRAF Phase 2 study, it will be unable to determine the correct size and mix of the CRAF fleet to meet future airlift requirements.

CONCLUSION

The nature of U.S. military operations in today's global environment requires DOD to be able to rapidly deploy personnel and cargo around the world and sustain forward deployed forces. DOD has taken a number of steps to strengthen the CRAF program while also ensuring that military aircrews receive required training. However, over the last few years, DOD has flown more hours than required to train its aircrews, thereby possibly reducing the level of peacetime business available to CRAF participants. The anticipated decline in DOD's peacetime business over the next few years, combined with continuing business pressures in a highly competitive industry, highlight the need for a process to ensure that DOD maximizes the use of its commercial partners. However, DOD does not use the process it has for monitoring training hours to determine when it can allocate eligible airlift missions to CRAF participants. If DOD does not use the information provided by its existing process, it will be unable to determine whether it is using commercial carriers to the maximum extent practicable, as required by DOD guidance. Further, DOD may be using its military fleet—which officials say is more expensive to operate than commercial alternatives—more than necessary, while risking the CRAF participation needed to ensure wartime readiness.

RECOMMENDATION FOR EXECUTIVE ACTION

To balance the use of military and civilian aircraft and ensure that commercial carriers participating in the CRAF program are used to the maximum extent practicable, we recommend that the Secretary of Defense direct the Secretary of the Air Force and the Commander, U.S. Transportation Command—in conjunction with the Commander, Air Mobility Command—to use the Air Mobility Command's existing process for monitoring training hours to determine when it can shift eligible peacetime airlift workload from military to commercial sources.

Agency Comments

We provided a draft of this report to DOD for comment. In its written comments, DOD concurred with our recommendation and stated that it believes implementing the recommendation will further improve the Civil Reserve Air Fleet program.

Zina D. Merritt
Director, Defense Capabilities and Management

Appendix I: Scope and Methodology

To determine whether DOD has been meeting its training requirements we reviewed Air Force guidance on the development of flying hour requirements, as well as DOD guidance on flying hours for training.[1] We also spoke with officials from U.S. Transportation Command (TRANSCOM) and Air Mobility Command (AMC) about how the training requirements are developed. We then analyzed flying hour data from the Reliability and Maintainability Information System (REMIS) to determine the extent to which the airlift fleet—C-5, C-17, and C-130—was being flown in excess of training requirements. We assessed the reliability of these data by interviewing officials from the REMIS program office at the Air Force Life Cycle Management Center to understand the steps that have been taken to ensure the reliability of the database. In addition, we reviewed documentation relating to the system and compared the data with alternate sources. We concluded that the data from REMIS were reliable for the purposes of this engagement. We then compared the flying hour data from REMIS to the flying hour requirements developed by AMC. To determine whether DOD was providing Civil Reserve Air Fleet (CRAF) participants with peacetime business, we interviewed officials with TRANSCOM and AMC on the management of the CRAF program, recent changes that had been made to the program, and concerns about the future of the program. In addition, we conducted interviews with representatives from 21 of the 30 CRAF participants who responded to our request for an interview in October and November 2012, to obtain information on the CRAF program, their perspective on what elements of the program worked and which did not, and their willingness to participate in the program in the near future. Since that time, 2 CRAF participants— one of which we interviewed—have gone out of business and are no longer members of the CRAF program. As of April 2013, there were 28 CRAF participants included in the CRAF program. We also analyzed program-related documents from TRANSCOM, AMC, and CRAF participants, as well as guidance on the use of CRAF and commercial transportation. Furthermore, we analyzed data from fiscal years 2001 through 2012 from two systems—the Commercial Operations Integrated System and an internal database managed by the Tanker Airlift Command Center (TACC) within AMC—to understand the extent to which CRAF participants are used compared with military airlift and foreign carriers. We assessed the reliability of these sources by reviewing documentation on to the systems, comparing these data with data from alternate sources and conducting interviews with knowledgeable officials. We concluded that the data from these systems were reliable for the purposes of this engagement.

To assess the extent to which DOD has justified restricting commercial carriers from transporting partial plane loads of cargo over channel routes, we reviewed DOD's policy for restricting commercial carriers from flying over channel routes.[2] The policy we reviewed helped us identify which channel routes were designated as restricted. We then conducted interviews with TRANSCOM and AMC officials to obtain information on the rationale for creating the policy and what operational and strategic benefits the policy provides for DOD. In addition, we reviewed fiscal year 2000 and fiscal year 2012 channel route airlift transportation data to determine the extent to which DOD was using military aircraft rather than CRAF participants to conduct channel route missions, and we discussed the circumstances surrounding those decisions with TRANSCOM officials. We also conducted interviews and obtained written responses from CRAF participants to obtain additional perspectives on how the policy is affecting the CRAF program. We also reviewed previously written reports and studies conducted by the RAND Corporation and the Council for Logistics Research Inc. that addressed DOD's use of channel routes, the impact of utilizing commercial carriers in lieu of military aircraft on DOD's aircrew training program, and the impacts the policy has had on overall cargo management. Reviewing this historical information provided us with additional insight into DOD's justification for implementing the policy.

To assess whether DOD has established future requirements for the CRAF program and how the planned size of CRAF compares with those requirements, we obtained and reviewed various studies conducted by DOD to assess its strategic airlift capabilities, such as DOD's Mobility Requirements and Capabilities Study – 2016, and the AMC 2012 CRAF study. We also collected fiscal year 2011 through 2013 data documenting DOD's current inventory of CRAF aircraft and compared these data with DOD's current airlift requirements. In addition, we conducted interviews with TRANSCOM and AMC officials to determine what steps are being taken to establish future requirements and to gain their perspective on the challenges they expect to face as they continue to manage the CRAF program. We also reviewed a provision in the National Defense Authorization Act for Fiscal Year 2013 that requires DOD to conduct a new study of mobility capabilities and requirements. We discussed the status of the requirement with TRANSCOM and AMC officials to determine what time frames and milestones have been established to begin and complete this study. We also reviewed DOD's defense strategic guidance issued in January 2012 to assess factors that may affect DOD's future airlift needs.

To gather information for these objectives, we reviewed documentation and interviewed officials from the following organizations:

DOD

- The Office of the Under Secretary of Defense for Acquisition, Technology, and Logistics
    - Office of the Deputy Assistant Secretary of Defense (Transportation Policy)
- U.S. Transportation Command
    - J-3 Operations and Plans, Sustainment Division (TCJ3-G)
    - Strategy, Policy, and Logistics (TCJ5/4)
    - Acquisition (TCAQ)
    - Office of the Staff Judge Advocate (TCJA)
    - Enterprise Readiness Center (ERC)
- Air Mobility Command
    - 618th Air and Space Operations Center (TACC)
    - Commercial Airlift Division (A3B)

Private Sector CRAF Participants

- See appendix III for the CRAF participants we interviewed
- National Air Cargo Association (NACA)

We conducted this performance audit from August 2012 to June 2013 in accordance with generally accepted government auditing standards. Those standards require that we plan and perform the audit to obtain sufficient, appropriate evidence to provide a reasonable basis for our findings and conclusions based on our audit objectives. We believe that the evidence and data obtained was sufficiently reliable for our purposes, and provides a reasonable basis for our findings and conclusions based on our audit objectives.

APPENDIX II: U.S. TRANSPORTATION COMMAND RESTRICTED ROUTE POLICY

MEMORANDUM FOR: International Carriers
Attachment 1
International Air Tender Restricted Routes
1 May 2012

Continued Policy Position 1 May 2012

As a condition of USTC approval of the International Heavyweight Air Tenders, Transportation Service Providers (Carriers/Agents/Forwarders) agree that they will not solicit or perform International Air Tender transportation services for Department of Defense consignors over validated OCONUS channel routes designated as "Restricted Routes" under USTC tenders.

Carriers are authorized to perform International Air Tender movement on Restricted Routes via "Spot Quotes" (or similar arrangements) for the DOD once the Transportation Officer has received/completed the requirements annotated in the DTR 4500.9R, Part II, Chapter 202, Paragraph Y.3.a.

International Routes	
United States	Afghanistan
United States	Germany
United States	Guam
United States	Italy
United States	Spain
United States	Japan
United States	Okinawa
United States	Korea
United States	Kuwait
United States	Bahrain
United States	Qatar
United States	Turkey
United States	United Kingdom
Germany	Afghanistan
Germany	Italy
Germany	Turkey
Germany	Kuwait
Germany	Qatar
All	Diego Garcia BIOT #
All	Guantanamo Bay, Cuba #

CONUS Routes	
United States	Alaska
United States	Hawaii

= No Commercial lift available to these areas

Notes: United States includes all 50 states. Restriction applies both ways for all route pairs.

OPR: USAF HQ AMC/A4TC Aerial Port Lane (DSN 779-4434 Comm 618-229-4434)
Partnered with USTC/J3-G/J3-S, USTC/J5/4-LD, USTC/AC
Coordinated by USTC TCAQ-CI/ dorothy.harris@ustranscom.mil (DSN 770-7068 Comm 618-220-7068)

APPENDIX III: LIST OF CIVIL RESERVE AIR FLEET PARTICIPANTS (AS OF APRIL 2013)

CRAF Carrier	Interviewed
1. ABX Air, Inc.	√
2. Air Transport International LLC	√
3. Alaska Airlines, Inc.	√
4. Allegiant Air LLC	
5. American Airlines, Inc.	
6. Atlas Air, Inc.	√
7. Delta Air Lines, Inc.	√
8. Evergreen International Airlines, Inc.	√
9. Federal Express Corp.	√
10. Hawaiian Airlines, Inc.	
11. Jet Blue Airways Corp.	√
12. Kalitta Air LLC	
13. Lynden Air Cargo LLC	√
14. Miami Air International, Inc.	√
15. MN Airlines LLC (DBA Sun Country Airlines)	√
16. National Air Cargo Group, Inc.(DBA Murray DBA National Airlines)	√
17. North American Airlines, Inc.	
18. Northern Air Cargo	
19. Omni Air International, Inc.	√
0. Polar Air Cargo Worldwide, Inc.	√
21. Ryan International Airlines, Inc.*	√
22. Sky Lease 1, Inc. (DBA Trade Winds Airlines)	√
23. Southern Air, Inc.	√
24. Southwest Airlines Company	√
25. Tatonduk Outfitters, Ltd. (DBA Everts Air Cargo)	√
26. United Airlines, Inc.	√
27. United Parcel Service Company	√
28. US Airways, Inc.	
29. World Airways, Inc.	

Source: GAO.

Note: Ryan International Airlines, Inc which previously declared bankruptcy has gone out of business since our meeting and was not a CRAF participant in April 2013.

End Notes

[1] Aircraft committed to the CRAF program can be activated during a war-time or emergency scenario under a stage I, II, or III CRAF activation. In exchange for participation in the program, carriers are eligible to fly peacetime airlift missions under a non-activated status.

[2] DOD refers to any business performed by CRAF participants outside of an activation as commercial augmentation, otherwise referred to as "peacetime business" throughout this report. DOD's last CRAF activation occurred from February to June 2003 at the start of Operation Iraqi Freedom.

[3] *See* National Security Decision Directive 280, National Airlift Policy (June 24, 1987) (hereinafter cited as National Airlift Policy (June 24, 1987)). As described by the policy, the national defense airlift objective is to ensure that military and civil airlift resources will be able to meet defense mobilization and deployment requirements in support of U.S. defense and foreign policies. *See id.*

[4] *See* Department of Defense Instruction 4500.57, Transportation and Traffic Management (Mar. 18, 2008) (hereinafter cited as DODI 4500.57 (Mar. 18, 2008)).

[5] GAO, *Military Airlift: DOD Should Take Steps to Strengthen Management of the Civil Reserve Air Fleet Program*, GAO-09-625 (Washington D.C.: Sept.30, 2009).

[6] Institute for Defense Analysis: *Civil Reserve Air Fleet: Economics and Strategy*, IDA paper P-4373; August 22, 2008; Council for Logistics Research, *Civil Reserve Air Fleet Study Report*, July 31, 2008.

[7] *See* H.R. Rep. No. 112-479, at129-30 (2012).

[8] We obtained training requirement data on DOD's flying hour program from the Air Mobility Command.

[9] *See* National Airlift Policy (June 24, 1987).

[10] DOD defines the mobilization base as the total of all resources available, or that can be made available, to meet foreseeable wartime needs. Such resources include the manpower and materiel resources and services required for the support of essential military, civilian, and survival activities, as well as the elements affecting their state of readiness. Joint Chiefs of Staff, Joint Pub. 1-02, Department of Defense Dictionary of Military and Associated Terms, at 189 (Nov. 8, 2010) (as amended through Mar. 15, 2013).

[11] The Commercial Airlift Review Board is comprised of DOD senior officials and makes decisions regarding the use or nonuse of air carriers participating or attempting to participate in the DOD air transportation program.

[12] To incentivize participation in the CRAF program, DOD awards CRAF carriers mobilization value points. Upon commitment and acceptance to the CRAF program, the CRAF carrier's aircraft are assigned mobilization value points based on factors including range, payload, and productive utilization rate, essentially determining the value of each aircraft to the program. These points are used to determine the amount of peacetime business CRAF carriers will be awarded.

[13] *See* 49 U.S.C. § 41106. The Fly CRAF Act applies to the transportation of passengers or property by CRAF-eligible aircraft obtained by DOD through a contract for airlift services. *See* § 41106(a)-(c). CRAF-eligible aircraft are aircraft of a type the Secretary of Defense has determined to be eligible to participate in the CRAF program. *See* § 41106(e). Subject to carrier availability, such transportation may be provided only by carriers that hold a certificate issued under section 41102 and either have aircraft in CRAF or offer to place aircraft in CRAF. *See* § 41106(a)-(c).

[14] Even where a CRAF carrier is unavailable, DOD may be required to use an available U.S. carrier pursuant to the Fly America Act. *See* 49 U.S.C. § 40118.

[15] Oversized cargo is air cargo that exceeds the dimensions of a standard (463L) pallet, but is air transportable on military and some civilian or foreign cargo carriers. Outsized cargo exceeds the dimensions of oversized cargo and requires the use of a military C-5 or C-17 aircraft, foreign owned capable aircraft, or surface transportation.

[16] For example, DOD officials pointed to Defense Federal Acquisition Regulation Supplement section 247.301-71, which applies to supply contracts that will include a significant requirement for transportation of items outside the continental United States. In these circumstances, the source selection evaluation criteria for award of the supply contract should include an evaluation factor or subfactor that favors suppliers, third-party logistics providers, and integrated logistics managers that commit to using carriers that participate in certain readiness programs, including CRAF. *See* Memorandum from the Under Secretary of Defense for Acquisition, Technology, and Logistics, Transportation within Supply Contracts (July 28, 2011) (citing 48 C.F.R. (DFARS) § 247.301-71).

[17] *See* National Airlift Policy (June 24, 1987).

[18] DODI 4500.57, paras. 4.4, 4.4.4 (Mar. 18, 2008).

[19] A "revolving fund" is a fund established to finance a cycle of businesslike operations through amounts received by the fund. A revolving fund charges for the sale of products or services and uses the proceeds to finance its spending, usually on a self-sustaining basis.

[20] TRANSCOM, *United States Transportation Command 2012 Annual Report.*

[21] GAO, Military Airlift: DOD Should Take Steps to Strengthen Management of the Civil Reserve Air Fleet Program, GAO-09-625 (Washington, D.C.: Sept. 30, 2009).

[22] To maximize the value of entitlements, CRAF participants form teaming arrangements, which are created and managed by the participants themselves. The teams are comprised of a mix of passenger and cargo carriers that join together to pool their entitlements to DOD business; that is, the entitlement directly associated with a carrier's individual commitment is combined with the entitlements earned by other carriers on its team. DOD assigns peacetime business to the team based on the team's total entitlement and availability.

[23] A "load plan" is the plan for how to pack a shipment of cargo so that it will fit safely onto a given aircraft.

[24] *See* Memorandum For: International Carriers, Attachment 1, International Air Tender Restricted Routes (May 1, 2012). The memorandum provides that "[a]s a condition of USTC [TRANSCOM] approval of the International Heavyweight Air Tenders, Transportation Service Providers (Carriers/Agents/Forwarders) agree that they will not solicit or perform International Air Tender transportation services for Department of Defense consignors over validated OCONUS channel routes designated as 'Restricted Routes' under USTC tenders." However, the memorandum also provides that carriers are authorized to perform such movement on restricted routes through spot quotes or similar arrangements, once certain requirements have been met. For a copy of TRANSCOM's restricted route policy memo see appendix II of this report.

[25] *See* National Airlift Policy (June 24, 1987). As described by the policy, the national defense airlift objective is to ensure that military and civil airlift resources will be able to meet defense mobilization and deployment requirements in support of U.S. defense and foreign policies. *See id.*

[26] *See* DODI 4500.57, para. 4.4.4 (Mar. 18, 2008).

[27] Brian G. Chow, *The Peacetime Tempo of Air Mobility Operations: Meeting Demand and Maintaining Readiness* (Santa Monica, CA: The RAND Corporation, 2003).

[28] Brian G. Chow, *The Peacetime Tempo of Air Mobility Operations: Meeting Demand and Maintaining Readiness* (Santa Monica, CA: The RAND Corporation, 2003)

[29] *See* Pub. L. No. 112-239, § 141(c) (2013).

[30] DOD, *Sustaining U.S. Global Leadership: Priorities for 21st Century Defense*. (Washington, D.C.: Jan. 3, 2012).

[31] *See* § 141(c). Specifically, the provision requires DOD to conduct a study that assesses the end-to-end, full-spectrum mobility requirements for all aspects of the National Military Strategy derived from the National Defense Strategy that is a result of the 2012 Defense Strategic Guidance and other DOD planning documents. *See* § 141(c)(1).

[32] *See* § 141(c)(2)(B), (E), (G).

[33] Block hours include not only actual flight time, but also taxi time before take-off and after a landing.

End Note for Appendix I

[1] *See, e.g.*, Air Force Instruction 11-102, Flying Hour Program Management (Aug. 30, 2011); DODI 4500.57, para. 4.4.4 (Mar. 18, 2008); Air Force Instruction 11-101, Management Reports On The Flying Hour Program (Nov. 1, 2002) (certified current Dec. 31, 2009).

[2] *See, e.g.*, Memorandum For: International Carriers, Attachment 1, International Air Tender Restricted Routes (May 1, 2012).

INDEX

A

B

C

J

K

L

M

N

O

P

R

S

T